Spectrum Wars

The Rise of 5G and Beyond

For a listing of recent titles in the
Artech House Mobile Communications Library,
turn to the back of this book.

Spectrum Wars

The Rise of 5G and Beyond

Jennifer A. Manner

ARTECH HOUSE
BOSTON | LONDON
artechhouse.com

Library of Congress Cataloging-in-Publication Data
A catalog record for this book is available from the U.S. Library of Congress.

British Library Cataloguing in Publication Data
A catalog record for this book is available from the British Library.

ISBN-13: 978-1-63081-916-3

Cover design by Charlene Stevens

685 Canton Street
Norwood, MA 02062

Disclaimer: The views in this book reflect those of the author and do not necessarily reflect those of EchoStar/Hughes.

10 9 8 7 6 5 4 3 2 1

To Dr. Eric Glasgow,
who I love to the moon and back

Contents

4

5G Technology Overview 81

5

Drivers of Change, Use Cases Driving 5G, Future Challenges, and the Race to Be First 99

Foreword

I got to know Jennifer Manner when we both worked at the U.S. Federal Communications Commission (FCC). I was the Chief Technologist, and she was the Deputy Chief of the Public Safety and Homeland Security Bureau. We collaborated on a chapter of the U.S. National Broadband Plan that urged Congress to fund the creation of a nationwide network for first responders, such as firefighters, police, and paramedics, using an approach very different from anything the U.S. had ever done before. When we first met, I didn't realize that she was the author of *Spectrum Wars,* which already sat on my office shelf, or that she had played an important role in crafting spectrum policy as a Senior Vice President at Echostar. I did realize very quickly that she had extensive knowledge of the wireless world well beyond public safety, and she was great fun to argue with. Congress accepted the core of those recommendations, initially with $7 billion dollars, which eventually led to the nationwide deployment of a network that is now known as

FirstNet. Americans are safer as a result. Long-term vision does matter.

Luckily for us, Jennifer has now turned her attention to fifth-generation cellular (5G), a subject that is desperately in need of clarity. Even for people outside of the wireless industries, it's been hard to miss the loud fanfare surrounding 5G's arrival. As consumers, we've been inundated with breathless advertisements from cellular operators about this new service that we apparently cannot live without. Some economic studies have predicted that 5G will add more than $13 trillion to global economic output, which is almost a year of GDP for the entire European Union. Government leaders have touted proposed legislation and executive orders, sometimes suggesting that falling behind rival nations in the race to 5G could only bring catastrophe. This seems more reasonable when you consider the awe-inspiring use cases we often hear about, such as remote surgery, self-driving cars, ubiquitous sensors, and virtual reality. Should all these claims be believed?

Underlying these use cases are technical claims for 5G that are even more extraordinary, more so than many people realize. Increasing data rates to each device by a factor of 10 to 100 and making those high data rates ubiquitously available is only a start. One goal is to reach a reliability of 99.999%, so safety-critical applications like remote surgery can be supported over 5G. That is far more reliable than electric power to U.S. households, which means 5G cellular would be far more reliable than most home broadband services, since home broadband generally depends on electric power from the grid. After years of dropped connections, it's hard to imagine a cellular service that is more reliable than fiber. Another goal for 5G is to support a million devices per square km, which would be a big leap forward for the Internet of Things. That's like asking everyone in the state of Florida to grab their favorite wireless device, squeeze into a square region that is less than 3 miles (5 km) on each side, and simultaneously expect good cellular connectivity. Yet another 5G goal is to achieve a round-trip latency of 1 ms, to support highly interactive wireless services. 1 ms is the round-trip time for light traveling through a fiberoptic cable that is a mere 62 miles (100 km) long, so 1 ms is better than any latency that can ever be achieved when communicating with a server that is only a couple hours' drive away. One more important but often underappreciated goal is to drive down energy consumption per bit carried, along with operating costs.

However, you can find skeptics who do not expect all of these alleged benefits of 5G to be realized. Indeed, some believe that in the end, 5G will be little more than a minor upgrade between the more consequential 4G and 6G cellular generations. Personally, I sympathize a bit with the skeptics when I hear the typical claims about 5G and remote surgery. I can certainly imagine that wireless infrastructure deployed on hospital grounds will become important for daily medical functions, including surgery, but I wouldn't want my surgeon cutting into me via a typical consumer's 5G cellular service any more than I'd want that surgeon balancing on a unicycle throughout my operation. Even if it's theoretically possible, there are simpler, cheaper, and, most importantly, safer ways to achieve the same results. Yet marketing departments love to pitch remote surgery.

Reality will probably fall somewhere between the industry's vision and that of the skeptics. Many people assume we can tell from today's standard alone, but 5G capabilities depend not just on the standard itself (which probably won't be completed before 2022), or even on the equipment built from that standard, but also on how cellular operators (and perhaps other kinds of actors) actually deploy 5G equipment. It also depends on the extent to which spectrum in diverse frequency bands is available for 5G. Thus, the future of 5G depends on many decisions by government and industry leaders yet to come, including decisions about spectrum policy. To make the right decisions, we first need to understand the issues.

Jennifer Manner is the right guide through these issues. In addition to serving at the FCC as Deputy Chief of two different bureaus and various other capacities, she has led the spectrum policy team for two satellite companies, giving her an opportunity to observe spectrum policy all over the world. She has served on the U.S. delegations to a number of international treaty organizations, whether it's at a World Radio Conference, or a meeting of the International Telecommunications Union (ITU), which is the branch of the United Nations that is responsible for global spectrum policy. She has also debated spectrum policy at domestic regulators in countless nations. This international perspective yields an unusual strength of the book. Readers get a glimpse of how issues are handled at the ITU, the FCC, and other domestic regulators.

Jennifer is also able to craft a story that is useful to both experts and relative newcomers. I saw this at the FCC when we presented our vision to stakeholders, and the public. You can see it when she testifies before Congress, or teaches a class at Georgetown University. Perhaps this is why she is probably the only spectrum aficionado who produces prize-winning films that are shown at the Cannes Festival.

So, read this book. Learn what is driving 5G cellular forward, how the new generation of cellular infrastructure will fit with everything from emerging satellite systems to next-generation Wi-Fi, and what government and industry leaders might do domestically and internationally to accelerate progress. Do it quickly, before we must all confront the many mysteries of 6G cellular—whatever that turns out to be.

Jon M. Peha
Professor at Carnegie Mellon University,
and former Chief Technologist of
the Federal Communications Commission
December 2021

Preface

As the world's reliance on wireless technology increases, there is one certainty regardless of the generation of technology—there will be an increased demand to access the radiocommunications spectrum resource. And due to this demand, it is increasingly important for the world of spectrum management (domestic and international) to be understood for 5G and beyond wireless technologies. This is the genesis of this book.

The role of the spectrum manager and the actions they take has become increasingly important due to the repercussions spectrum resource use has on our daily lives. For example, in instances of harmful interference, there can be significant outages that impact people's ability to email and text, cutting us off from critical public safety communications such as emergency calling. In addition, to realize the benefits of new wireless technologies, it is the spectrum managers who identify and provide use of the spectrum resource, as well as develop the appropriate technical, operational, and licensing rules.

I hope this book will provide a greater understanding for people interested in our wireless world of how the spectrum resource is used and how decisions are made. I have seen many changes over the years and believe the greatest challenge we face in the 5G and beyond world is to ensure there is an efficient use of the spectrum resource. I hope this book will help address this and similar challenges.

This book would not be possible without the help of a number of people. First, I would like to thank my research assistant, Endeliza Hampton, who is a third-year law student at the University of Nebraska School of Law, who challenged herself, as well as me, as she helped put together this book. Endeliza' s help has been invaluable and I look forward to seeing what the future holds for her.

I also want to thank my dear friend and colleague, Donna Bethea Murphy. Donna and I began our careers together at the Federal Communications Commission and throughout our careers I have been very lucky to engage with Donna in, at times, daily discussions on the spectrum issues that we have faced during our careers. I credit these discussions for helping me work through some of the most difficult spectrum challenges I have had to face while in government and the private sector.

I also would like to thank several colleagues whom I have enjoyed having discussions and debates with about spectrum management issues. These include Dr. Jon Peha of Carnegie Mellon University, Dale Hatfield of the University of Colorado at Boulder, Kathleen Abernathy, Anders Johnson, Dr. Stagg Newman, Dr. Whitney Lohmeyer of Olin College of Engineering, Charlene King, and Dr. J. Pierre De Vries of the University of Colorado at Boulder. Each of these people, and so many more I am sure I am forgetting, make my role as a spectrum manager that much more interesting and meaningful. While I have benefited significantly from these discussions, the views expressed in this book are solely my own and do not necessarily reflect the views of Echostar Corporation.

I would also like to thank my dear husband, Dr. Eric Glasgow. Besides always being supportive and loving, Eric has spent much of our time together, whether on runs or just having dinner, listening and engaging with me as I struggle to solve the spectrum management issues that I have faced on a day-to-day basis. While I am sure Eric, who is the director of the Zebrafish Resource Center, and a professor at Georgetown University School of Medicine, would have preferred to talk about science, he has always been willing to spend the time with me to work through these complex spectrum issues. I cannot say enough about how much this means to me. And a final thank you to my pup, Charliedoodle, who is never too busy to listen to me work through a spectrum problem whether on a run or when just cuddling.

1

Introduction: The Changing World of Spectrum

1.1 5G and Beyond Technologies: Their Demand for Access to Spectrum

It has been almost two decades since my book, *Spectrum Wars,* was published at a time when the world was focused on deploying 3G, before 4G was even developed [1]. During the intervening years, the world of spectrum management has evolved to reflect the growing and changing needs of spectrum users. This includes the development of 4G and 5G, the need for access to more spectrum to support the users need for greater capacity and speeds, and the ability of new technologies to allow more efficient use of the spectrum resource [2].[1] This has resulted in spectrum manag-

1. For instance, the COVID-19 pandemic resulted in increasing user needs for access to wireless technologies to support internet and other users.

ers utilizing methods such as dynamic spectrum access and innovative auctions to allow greater access to the spectrum resource. This book explores the changes of the past two decades and how, as we embrace 5G and beyond technologies, we are responding to some of the greatest spectrum management challenges the world has ever faced.

5G and beyond telecommunications technologies hold great promise for the world; far greater than we were able to achieve with previous generations of wireless technologies. 5G is planned and, if successful, is on its way to create a highly connected world that brings highly reliable, fast, low latency, mobile and fixed telecommunications services to users wherever they may be. 5G is the first generation of wireless that includes all technologies, whether fixed and mobile terrestrial or fixed and mobile satellite services. And 6G promises connections beyond the physical world in what has been called the *Internet of Senses* opening a world of new possibilities to users [3].

But the question is, how do we get there? There are many requirements, including the financial wherewithal, to construct and operate the telecommunications networks to support these technologies along with whether the technologies will be able to meet 5G requirements for things such as latency and speed. At least of equal importance is ensuring that the 5G network operators and service providers have access to a sufficient amount of the radio frequency spectrum to support the expected traffic demands for the 5G and future wireless networks. This is where this book steps in. In this book, we are focused on examining the spectrum resource and how it is being managed to enable its most efficient use to support the global deployment of 5G and beyond for all technologies. We will also examine some of the more innovative solutions for making additional spectrum available for 5G. As we will see, oftentimes there are battles between current and future spectrum users (and at times within each group) over the best use of the spectrum resource. Some of the solutions provide promise for greater shared access to the spectrum resource and the ability to obtain greater spectrum efficiencies. However, in many cases, there is a cost to such a result; something we will also examine.

As the twenty-first century is already demonstrating, telecommunications will continue to have a growing influence on how we live, work, learn, and socialize, including in the most rural portions of the world. To enable this demand, great in-roads have been made, ensuring advanced communications services are available globally. For instance, today people throughout the world have access to some form of satellite telecommunications capabilities. In addition, approximately 45% of people in the world have a smartphone, with that figure expected to grow approximately 8% to 9% in the next year [4]. Building on this success to date, it has been estimated that the mobile telecommunications industry will connect all the world's population by 2030 [5]. And wireless and other industries are estimating that 5G will bring greater access and increased telecommunications capabilities to people across the globe on a more expedited basis than previously seen [6].

5G communications, if it delivers as promised, will encompass all wireless technologies (terrestrial and nonterrestrial), and will bring greater speeds, connectivity, capacity, and resiliency while lowering latency for latency sensitive applications to users across the globe [7]. While deployment of 5G has already begun in some countries, like the United States, many others, such as Madagascar and Ethiopia, have not even started [8]. In fact, in many countries around the world, we can only expect to see 5G deployment in the future as these countries grapple with still deploying earlier generations of communications technology, such as 3G.

Ultimately, the promise for those who deploy 5G networks is that it will create an always-connected world for devices as well as everything else from work, socializing, and education to school, healthcare, government, finances, and so much more. Future technologies, such as 6G communications promise even more, having the potential to bring users even higher speed and reliability, as well as lower latency [9].

1.2 The Search for Access for Additional Spectrum to Meet 5G and Beyond Goals

To lay the framework for evaluating spectrum management for 5G and beyond, it is critical to understand the availability of spectrum today to support these uses, the role of technology, and the role of spectrum management. Below we examine the issue of spectrum availability today, the role of technology in freeing up more spectrum for 5G and beyond use and, of equal importance, the role of the regulator or spectrum manager in regulating this scarce resource.

1.2.1 Spectrum Availability Today

The use and management of the radio spectrum resource undergoes constant change as new technologies, including new generations of technologies, and new use cases are introduced, pressuring, and in some cases, forcing the regulators who manage the spectrum resource to respond. The latest strain on the spectrum resource is the development of 5G due to the large amounts of bandwidth needed to meet demand for greater capacity and speeds of wireless communications [10]. As 6G and other future generations of technology evolve, even greater demands for bandwidth will appear, further stressing the spectrum management regime.

The spectrum resource is already crowded with radio services allocated and assigned for use across the frequency bands for a wide range of radio services from mobile services to radio astronomy to the Earth exploration service and beyond [11]. Radio frequency-based telecommunications services can be defined as those using the electromagnetic waves of frequencies lower than 3000 GHz, propagated in space without artificial guide [12]. These radio uses, as we will discuss later, are called radio services. These are quite diverse ranging from *mobile services,* which allow you to access the internet or make a phone call on your smart phone, to using spectrum assigned to the *space services* and *radio navigation services*. Space services track the rockets that launch satellites and people into space, while location tracking technologies, such as GPS, fall into radio navigation network services. Yet each radio service operates pursuant to certain technical requirements such as power limits. To make the most efficient use of the radio spectrum resource, frequency bands are often shared with technically compatible services; however, such sharing may also impose technical constraints on the operation of these services to avoid harmful interference with one another.

Despite the apparent congestion of the spectrum resource, no one really knows the actual severity of congestion of the spectrum resource for several reasons. First, there is no international or national registry of what the spectrum resource is actually being used for. Even in heavily licensed bands, domestic regulators may not know the actual use of the licensed spectrum, including how much of the relevant band is used, and to what extent. This is compounded by the fact that no one has systematically reviewed every frequency band in the radio spectrum resource to determine what is actually in use. And even if such an analysis was performed, it is clear that even for the most intensely utilized bands, greater efficiencies of use could be made.

Further, an exhaustive search of the literature and research shows that there is no agreed way to calculate what the available capacity of the radio spectrum resource is. And, today, there is almost no usable spectrum that does not have some assigned use, even in the highest frequency bands. Accordingly, when spectrum managers look to give access to a new use or increase access to additional bands for a new use, they must look at potential individual bands they want to access to see if (1) there is congestion in the band, and if so, is the band utilized efficiently, and (2) what is the value of the incumbent uses and, if they are valuable, how can they best be accommodated (e.g., should they remain in the current band or is relocation appropriate).

Therefore, if one really wants to determine what spectrum is actually in use today and to what extent, it is critical to examine each frequency band (i.e., a subset of frequencies, such as 27.5–29.5 GHz) in a particular geographic region individually (i.e., global, regional, country-wide or a smaller service area, such as a county). This examination will look at the use or uses of the band (e.g., mobile telephony, fixed wireless broadband, GPS) and how efficient that use is in that region, as well as whether the service serves the entire geographic area. The answer to these questions will be an important factor in determining whether and how an individual band can be used globally or in a particular country/region for future uses or whether use can be expanded. However, because of the interest of incumbent users in retaining access to spectrum they already hold, such analyses are rarely conducted and, if they are, may not be accurate because of the unwillingness of the incumbent to cooperate for fear of losing access to parts or all of their spectrum resource.

Another important calculation in determining the availability of spectrum use of a frequency band is the perceived value of its current and future use by the incumbent services [13]. This is largely a subjective determination and may change over time depending on the needs of individual countries and users. For example, while wireless telegraph was an extremely valuable use of the spectrum in the first half of the twentieth century, today it would be seen as essentially a waste of the spectrum resource as new uses have replaced its utility. Further, for developed countries, like those in the European Union, the United States, and Canada, 5G may be a priority, but countries in the developing world may only be at the stage of deploying 3G technologies. This disparity in telecommunication technology deployment was demonstrated at the International Telecommunications Union's (ITU's) 2015 World Radiocommunication Conference (WRC-15) where additional spectrum was being identified for 5G (or what is called international mobile telephony [IMT] at the ITU). In response to a request for the 27.5–30 GHz band to be considered for IMT identification at WRC-23 because of its importance to meet spectrum needs, a delegate whose country utilized that band for satellite objected. In his objection, he made the point that his country was not focused on 5G because they were still working to have 2G available in their country. Ultimately, an overwhelming number of delegates chose not to include the examination of this band on the WRC-19 agenda. As this example shows, needs for access to the spectrum resource to support communications technologies may differ across countries and even regions.

In addition, when considering the use of a frequency band for a new use, the spectrum manager must also consider technically whether this frequency band would meet the technical requirements for this use as well as the impact on the incumbent user. Different portions of the radio frequency resource have different physical characteristics and are better for different types of communication [14]. For instance, the lower portions of the radio frequency band propagate further, which means they more easily pass through buildings and trees. For this reason, these frequency bands have traditionally been seen as valuable for mass communications services, such as cell phones, TV, and radio.

However, with 5G we are seeing for the first time a broader need for multiple types of spectrum: low-, mid-, and high-band spectrum (see Table 1.1). This is in part because of the differing needs of the 5G infrastructure and ecosystem to meet user demands. For instance, high-band frequencies are seen as an important resource for bringing high-speed services to dense urban areas because of the band's poor propagation characteristics. Accordingly, using a mix of frequency bands, 5G networks have the potential to leverage the spectrum most appropriate for the planned use to meet user demands.

Once this initial analysis is determined, the spectrum manager then must understand technically whether and how the new use could operate in the band, including the potential impact to services operating in-band and in adjacent bands. As discussed above, this determination will take into account the efficiency of the current use, the value of the current use, the value of the new use, and whether the relevant frequency band is appropriate for such use. Also considered is whether the proposed new use could result in harmful interference to the incumbent service and adjacent uses, and if so, what protections might be afforded to limit interferences or what other actions could be taken (e.g., relocation) [15].[2]

Later we will examine real-life examples of where regulators have made such choices. But, before we do, it is important to ex-

Table 1.1

Examples of Low-, Mid-, and High-Band Frequency Bands

Low-Band Frequency	Mid-Band Frequency	High-Band Frequency
1 to 2.6 GHz	2.6 GHz to 6 GHz	Above 24 GHz

2. The ITU Radio Regulations (RRs) define harmful interference as interference which endangers the functioning of a radio navigation service or other safety services or seriously degrades, obstructs, or repeatedly interrupts a radiocommunications service operating in accordance with the Radio Regulations.

plore how technology is enabling greater use of the spectrum resource than ever before.

1.2.2 The Role of Technology

As will be shown throughout this book, technology plays a key role, although not the only one, in determining how efficiently and how to best use the spectrum resource. In order to access sufficient spectrum to meet the needs for new and innovative 5G and beyond services to meet growing user demands, technical methods must be created and utilized that can access already used spectrum or open up higher frequency bands to more intensive use [16]. Technologists have responded to this important challenge by developing solutions to advance this effort.

As there is increasing pressure for greater access to the spectrum resource to deliver the latest wireless services (like 5G) there has been an increasing focus on developing new technologies to make more efficient use of the spectrum resource by industry and other spectrum users, such as governments. For example, today's commercial satellite systems have transitioned to use spot beams from transponders, which results in greater spectrum reuse, hence increasing efficiency [17]. Other recent developments include cognitive technologies to use the spectrum resource more efficiently to support the anticipated massive data requirements of end users in a 5G world [18].

Additionally, technologists continue to develop spectrum sharing tools, including enabling dynamic sharing, which allows more than one radio service or spectrum user to operate in the same frequency bands on a primary or secondary basis without causing harmful interference [19]. An example of dynamic spectrum sharing, which will be discussed later, is the U.S. Federal Communications Commission's (FCC) Consumer Broadband Radio Service (CBRS) regulatory regime in the 3.5 GHz band. CBRS uses a tiered structure to protect incumbents (the fixed satellite service [FSS] and government radars) while enabling sharing between new terrestrial wireless users on both an unlicensed and licensed basis [20]. CBRS spectrum is managed through a spectrum access system (SAS) that centrally coordinates access to the shared spectrum, enforcing priorities and modeling the radio frequency

(RF) environment [21]. Using this framework the FCC created a regime that protected incumbent government radar use and existing satellite use, while opening this same frequency band for use by 5G terrestrial mobile services on a licensed and unlicensed basis (respectively, priority access and general authorized access). The CBRS regulatory framework resulted in an FCC auction (FCC Auction 105), that earned $4.58 billion in gross revenues for the U.S. Treasury. Major players in the wireless industry, including T-Mobile, Verizon, and DISH Networks participated in the auction and won spending significant money, which demonstrates the faith of the marketplace in dynamic spectrum sharing, as well as the value of this spectrum for 5G [22].

An example of another sharing regime (not involving dynamic spectrum) is the sharing of spectrum that satellite operators have engaged in for years between FSS geostationary orbit (GSO) satellite networks [23]. For example, there is a long-standing policy for a 2-degree separation between FSS and GSO networks. This policy ensures equitable and efficient use of the orbital resource, but also efficiently shares the FSS spectrum among different FSS networks by limiting radiation from Earth stations toward adjacent satellites and downlink EIRP density of power-flux density generated by space stations.

Technologists have also been developing solutions to utilize higher frequency bands that until recently have had little demand and, accordingly, have largely remained unused. For example, millimeter-wave frequency bands (i.e., those bands above 24 GHz) have until recently not been considered to support ubiquitous services, such as terrestrial mobile service or FSS consumer broadband services, largely because of their propagation characteristics. However, the increasing demands for 5G and broadband services, has led to the development of cost-effective technologies to access many of these frequency bands for terrestrial use, including for small cells, which will allow high speed services to be delivered over relatively short distances (e.g., 1 km).

As previously noted, the terrestrial wireless industry is looking for low, middle, and high band spectrum to support their roll-out [24]. Therefore, use of the spectrum above 30-GHz bands, which is considered high band spectrum, is critical for the rollout plans of major mobile services operators such as Verizon, Vodafone, Telefonica, and AT&T [25]. Similarly, while many of the spectrum bands above 30 GHz (e.g., 40–42 GHz) have been available to the fixed satellite services for way over a decade, the recent evolution of satellite technology is now only allowing access to this band for both gateway and user terminal use, with systems such as HughesNet deploying satellite networks which operate in this frequency range.

However, the ability of multiple services to access additional spectrum, tied with increasing demand, means that often services where sharing is extremely difficult (i.e., where two different services are ubiquitously deployed), there is competition among potential users for primary access to the frequency band. This often requires spectrum managers to make difficult decisions, especially when relocation or curtailed use of the incumbent is the likely result. The millimeter wave bands provide a perfect example of this situation.

Since technology has allowed greater flexibility for the operation of different services, there has been a movement towards relocating existing users whose use of the spectrum is not as valued or as efficient as another use of the spectrum. Needless to say, such action is not without controversy with incumbents generally hostile to relocation. However, in some recent cases, where financial incentives are provided along with a path to continued operations, the incumbents have supported relocation of their services. A good example that we will examine further is with regard to the C band in the United States. In this case, the FCC was able to provide reimbursements and incentive payments to FSS operators who relocated their service to another band in order to allow the band to be auctioned for 5G mobile terrestrial use [26]. This, in essence, made these satellite operators whole as they received funding to buy replacement satellites that could operate in the spectrum where they were relocated without disrupting service to their end users.

1.2.3 The Role of the Spectrum Management and Regulatory Processes

Spectrum management is critical to understand current and future spectrum wars over how to best utilize the spectrum resource. In this regard, two important goals of spectrum resource are to ensure the efficient use of the spectrum resource [27] and avoiding harmful interference [28]. However, decisions on spectrum management are also influenced by technical, economic and, at times, political factors, as well as national security and other policy considerations [29].

As we will see throughout this book, the policy goals of the spectrum manager can be used to circumvent what may be a technical decision on how to utilize the spectrum resource in the name of achieving an important public policy. For example, recently, the FCC authorized increased flexibility for Ligado, a U.S. L-band mobile satellite operator with the rights to operate a terrestrial component, that some government and commercial users have strongly argued would lead to harmful interference into sensitive and critical communications networks, including GPS [30]. The Ligado decision created a technical framework that allows Ligado the ability to use low band spectrum around 1.6 GHz for 5G, while putting into place operational restrictions to protect incumbent in-band and adjacent-band services, including GPS [31]. Despite this effort by the FCC, there has been an outcry by U.S. government and commercial users, as well as politicians, that there will be harmful interference into critical GPS systems including those used by the U.S Department of Defense (DOD) [32]. In this case, the pressure has been so intense by the incumbent users, that the U.S. Congress passed a DOD authorization bill that does not allow the DOD to contract with a commercial entity that has rights to this spectrum, without a waiver from the Secretary of Defense [33], a very high barrier.

In other actions aimed at promoting 5G, some governments are building-out and operating 5G networks themselves. Despite, as discussed later, existing precedent that demonstrates that governments are largely inefficient operators of telecommunications networks [34].[3] For example, Malaysia has had a competitive commercial wireless industry for years, with 2020 reports showing

3. The basis for such models is likely to include lack of faith in the commercial marketplace as well as a need for control. Unfortunately, as time has shown us, the government-owned monopoly model will likely result in higher prices, less choice and innovation, and lower quality services.

Malaysia operating with several high-quality mobile wireless networks [35]. However, in 2021, the Malaysian government took a surprising decision to announce that it was going to build and operate a government-owned 5G network, with minimum private sector involvement [36]. It is unclear that such an approach makes the most efficient use of the spectrum resource.

We will also explore how other governments, such as some in Europe, are making spectrum available for 5G through auctions that contain conditions requiring build-out to rural areas for the provision of broadband and 5G services, in an effort to solve the digital divide. As we examine these and other models, we will explore whether they are likely to be successful and how this approach may impact other spectrum users as well as how likely they are to achieve their stated policy goals.

Overall, the introduction of 5G and future technologies places increasing pressure on the spectrum management regime to evolve to support this new generation of telecommunications ecosystem and network, which it is claimed requires an increasingly large amount of spectrum to be successful. The good news is there is strong economic and public policy pressures on all actors, including governments, operators, and users, to work together to find solutions to support additional spectrum use.

Nonetheless, we have never faced the introduction of technologies, such as 5G and beyond, that bring with them the need to access so much spectrum. As spectrum managers try to meet these demands it is important that they take a technology neutral approach[4] and balance the need of all technologies to have access to the spectrum required to meet their needs [37]. This is critical to ensure that a wide variety of technologies needed for different uses, as well as competitors, exist in the marketplace. However, as we will explore later, this is a very difficult task. Oftentimes balance cannot be achieved which may mean that some technologies suffer from lack of spectrum access, negatively impacting their ability to meet user demand.

Finally, it is important to recognize that as we enter the 6G world, we are also likely to increase demand for access to the

4. As discussed later, a balanced approach to technology neutrality does not require the equal distribution of the spectrum resource.

spectrum resource and other resources that are unregulated, such as optical technologies. We will explore this in Chapter 8 along with the goal for spectrum managers to anticipate and plan for these needs now and take appropriate actions to address upcoming issues.

1.3 Next Steps

We begin our discussion on the battles over the spectrum resource in the 5G and beyond world by establishing a baseline of spectrum regulation today and how we got here. Chapters 2 and 3 lay the groundwork for the domestic and international regulatory regimes surrounding the spectrum resource.

Chapter 2 specifically examines the role of the ITU, including its process and roles in managing the spectrum and orbital resources.

Chapter 3 supplements Chapter 2 by discussing the domestic regulatory process including how actions at the ITU and its World Radiocommunication Conferences are implemented domestically. In addition, Chapter 3 also delves into domestic licensing processes, which are critical to understand as a part of spectrum management, as well as the important role of spectrum harmonization.

Chapter 4 provides a concise technology primer of terrestrial and nonterrestrial technologies. On the wireless side we will examine a range of technologies from Wi-Fi to small cells. We will also explore the importance of low-, medium-, and high- band spectrum to be able to meet the needs of this new generation of wireless services. This chapter also examines the range of nonterrestrial technologies from satellite to drones, and other high altitude platforms that are planned to support 5G and beyond wireless services. This will include looking at the impact of mega-NGSO constellations and how they will impact the spectrum and interference environment, as well as exploring plans for space, including Nokia's recent announcement that is building a 5G network on the moon [38].

Chapter 5 will be a short dive into 5G use cases. This chapter, in particular, will touch on, traditional mobile wireless use cases, such as smartphone use, as well as the array of additional use cases we will see through 5G technology; from precision agriculture to machine-to-machine communications and more.

We will then shortly explore what is driving the demand for 5G. For example, what are the direct drivers for the need for such a reliable, high-speed service. Things like the growing need for autonomous vehicles to increase passenger safety, supply chain efficiencies, and precision agriculture which increases agriculture, reduces climate change, and conserves scare resources. We will also examine how the COVID-19 pandemic is shaping what 5G ultimately looks like as many of our daily habits of work, socializing, and education changed. And of course, our growing desire to have everything connected has led for increased needs for the Internet of Things (IoT) which requires access to harmonized frequency bands.

We will then look at the role of hybrid technologies, including networks and end user services that combine the benefits of both GSO and nongeostationary orbit (NGSO) satellite networks or combine terrestrial wireless with NGSO satellite networks, sharing the same spectrum. We will also look at the role of multiple technologies working together or on a complimentary basis to increase the end user experience or reach additional users, such as in rural areas. A good example of this is the use of fixed service microwave or satellite to provide backhaul for a terrestrial mobile 5G system. Of course, successful hybrid or network of networks systems require harmonized standards that enable these systems to seamlessly interoperate, so we will touch on the standards process here as well.

Following this discussion, we will spend some time examining how we meet growing spectrum demand. In this section, we will look at the changing world of spectrum management including looking at the tools that are becoming increasing prevalent to make additional access to 5G spectrum a reality; technologies from dynamic spectrum access, such as for CBRS, as we discussed above, to the use of government systems, to the role of unlicensed or license exempt spectrum.

We then discuss the methods available today to get spectrum out into the marketplace. This area is perhaps the most challenging for regulators, users, technologists, and others since the legacy approach to spectrum assignment and allocation means that today there is no unburdened portion of the spectrum resource. We will discuss how regulators address this issue, and who the winners and losers are in this very real, costly, and complex battlefield. This includes examining different approaches to clearing, sharing mechanisms, the range of auction approaches we are seeing from incentive auctions to those that impose social obligations, to secondary markets and more. We will also explore how government spectrum is becoming increasingly at risk as governments understand the economic and other benefits surrounding the use of 5G networks and beyond. Next we will delve into how spectrum is being made available in unique ways for nonterrestrial technologies from NGSOs to drones as well as in deep space and how these uses may impact use of the spectrum resource on Earth through, for instance, aggregate interference.

Before we conclude, we will explore 6G and beyond technologies. As we will discuss, 6G poses a range of unprecedented challenges for spectrum managers as the world looks to create an Internet of Sensors. While it is still too early to tell, we expect that 6G systems will enable us to move from gigabit to terabit capabilities, as well as to submillisecond response times [39]. 6G will enable new applications such as real-time automation or extended reality sensing (Internet of Senses), collecting data that will enable us to make a digital twin of the physical world [40]. This is driving, to some extent, looking to use substitutes or additions to radio spectrum to deliver communications services wirelessly to end users, such as optical technologies. In this chapter, we will make some recommendations to regulators on spectrum management as they plan for this evolving wireless world.

Finally, we conclude our journey by looking at the major challenges that will need to be addressed by spectrum mangers in the 5G and beyond world of spectrum management. These challenges include:

- Increasing the use of the spectrum resource;
- Ensuring the protection of incumbent users;
- Making additional spectrum available on a technology neutral basis;

- Reducing regulatory burdens and fees;
- Achieving global harmonization;
- Increasing competition.

We are at an exciting time in the history of telecommunications and the management of the spectrum resource. This excitement is compounded by the fact that for the first time, 5G and even more dramatically, 6G, networks will comprise a variety of terrestrial and nonterrestrial communications technologies, increasing the user experience. By exploring the regulatory structures, the needs for 5G and beyond, and the tools at our disposal, we will be in a better position to meet the challenges we face.

References

[1] Manner, J., *Spectrums Wars: The Policy and Technology Debate*. Boston: Artech House, 2003, 186 pages.

[2] "The Promise of 5G", PWC, accessed June 07, 2021, https://www.pwc.com/us/en/services/consulting/library/consumer-intelligence-series/promise-5g.html.

[3] Fisher, T., "6G When It Is and When to Expect It." *Lifewire*, Feb. 19. 2021. Accessed June 02, 2021. https://www.lifewire.com/6g-wireless-4685524.

[4] "How Many People Have Smartphones in 2020," *Oberlo*, accessed June 07, 2021, https://www.oberlo.com/statistics/how-many-people-have-smartphones.

[5] Barrett, J., "Global Opportunities for 5G Terrestrial Communications." *ITU News*. August 19, 2019. Accessed June 02, 2021. https://news.itu.int/global-opportunities-for-5g-terrestrial-communications/available.

[6] Id; see also EMEA Satellite Operators Association. "Supporting 5G Services." ESOA. Accessed June 02, 2021. https://www.esoa.net/5g/supporting-5g-services.asp.

[7] https://www.cnn.com/interactive/2020/03/business/what-is-5g/index.html. EMEA Satellite Operators Association. "Satellite Communications: An Integral Part of the 5G Infrastructure." ESOA, July 2020. https://esoa.net/cms-data/positions/2278%20ESOA%205G%20Ecosystems%20UPDATE%20NOV%202020.pdf.

[8] Husami, K., "Madagascar telco instructed to put 5G plans on hold," Inside Telecom, July 28, 2020, https://www.insidetelecom.com/madagascar-telco-instructed-to-put-5g-plans-on-hold.

[9] "What is 6G? - What Can We Expect from It?" Accessed June 02, 2021. https://whatis6g.com/definition.

[10] Bazelon, Coleman, McHenry, and Giulia. "Spectrum Value," *Telecommunications Policy* 37, no. Issue 9 (2013): 737. Lenzel, L., "Understanding Solutions For The Crowded Electromagnetic Frequency Spectrum." Electronic Design. March 21, 2012. Accessed June 02, 2021.

[11] https://www.electronicdesign.com/technologies/communications/article/21799723/understanding-solutions-for-the-crowded-electromagnetic-frequency-spectrum.

[12] Wikipedia. 2021. "Radio Spectrum." accessed June 09, 2021, https://en.wikipedia.org/wiki/Radio_spectrum.

[13] ITU, ""Exploring the Value and Economic Valuation of the Spectrum Resource," accessed June 07, 2021, https://www.itu.int/ITU-D/treg/broadband/ITU-BB-Reports_SpectrumValue.pdf.

[14] Dunbar, B., "Characteristics of Electromagnetic Spectrum." NASA, Feb. 28, 2018. Accessed June 02, 2021. https://www.nasa.gov/directorates/heo/scan/spectrum/txt_characteristics_spectrum.html.

[15] ITU RR No. 1.169.

[16] Lazarus, M., "What 5G Engineers Can Learn from Radio Interference's Troubled Past." IEEE Spectrum, June 9, 2016. Accessed June 02, 2021. https://spectrum.ieee.org/telecom/wireless/what-5g-engineers-can-learn-from-radio-interferences-troubled-past.

[17] Gannon, P., "Understanding the Basics of HTS." Businesscom Networks, March 1, 2018. Accessed June 02, 2021. https://www.bcsatellite.net/blog/understanding-the-basics-of-hts/.

[18] Nathan, S., "World Records in 5G Wireless Spectrum Efficiency Using Massive MIMO." The Engineer, December 11, 2017. Accessed June 02, 2021. https://www.theengineer.co.uk/5g-wireless-spectral-efficiency-using-massive-mimo.

[19] Ericsson, "Breakthrough 5G data call using dynamic spectrum sharing to accelerate nationwide 5G deployments, available at Dynamic spectrum sharing for 5G deployments." Accessed June 02, 2021, https://www.ericsson.com/en/news/2019/9/ericsson-spectrum-sharing.

[20] Federal Communications Commission, "Citizens Band Radio Service (CBRS)," accessed June 03, 2021, https://www.fcc.gov/wireless/bureau-divisions/mobility-division/citizens-band-radio-service-cbrs.

[21] Metaswitch, "What is the Citizens Broadband Radio Service (CBRS)?" accessed June 02, 2021, https://www.metaswitch.com/knowledge-center/reference/what-is-the-citizens-broadband-radio-service-cbrs.

[22] Federal Communications Commission, "Auction 105: 3.5 GHz Band," accessed June 03, 2021, https://www.fcc.gov/auction/105.; Wagner, Alex, FCC Announces Results of CBRS Spectrum Action that T-Mobile Participated In," T-Mo Blog Sept. 02, 2020, https://www.tmonews.com/2020/09/fcc-cbrs-spectrum-auction-results-t-mobile-verizon-dish.

[23] Federal Communication Commission, "Sharing of Spectrum Used by Satellite Systems," Jose Albuquerque, NSMA, May 17. 2017, accessed June 03, 2021, https://nsma.org/wp-content/uploads/2017/05/2017-15-nsma-sharing-of-spectrum-used-by-satellite-systems.pdf.

[24] Heinzman, A., "Not All 5G Is Equal: Millimeter Wave, Low-Band and Mid-Band Explained," How To Geek, July 23, 2019. Accessed June 02, 2021, https://www.howtogeek.com/428337/not-all-5g-is-equal-millimeter-wave-low-band-and-mid-band-explained.; Nokia, "5G Spectrum Bands Explained," accessed June 02, 2021, https://www.nokia.com/networks/insights/spectrum-bands-5g-world.

[25] Sha, A., "List of 5G Bands in the US for Verizon, AT&T, Sprint and T-Mobile," Beebom, January 21, 2021, accessed June 02, 2021, https://beebom.com/list-5g-bands-us-verizon-att-sprint-t-mobile.

[26] See, e.g., Jewett, Rachel, "Satellite Operators Agree to Clear C-Band on FCC's Timeline," Satellite Today, June 1, 2020, https://www.satellitetoday.com/5g/2020/06/01/satellite-operators-agree-to-clear-c-band-on-fccs-timeline.

[27] U.S. Department of Transportation, "What is Spectrum Management," Washington, DC: DOT, accessed June 03, 2021, https://www.transportation.gov/pnt/what-spectrum-management.

[28] Manner, J., "Spectrum Interference in the New World of Space and 5G," Via Satellite, accessed June 03, 2021, https://www.satellitetoday.com/innovation/2019/01/18/spectrum-interference-in-the-new-world-of-space-and-5g.

[29] See, e.g., Stephen, et. al, Innovations in Spectrum Management: Enabling community networks and small operators to connect the unconnected," Internet Society, accessed June 02, 2021, https://www.internetsociety.org/resources/doc/2019/innovations-in-spectrum-management.

[30] XYHT, "Running Interference," accessed June 10, 2021, https://www.xyht.com/gnsslocation-tech/running-interference-ligado.

[31] In The Matter of Lightsquared Technical Working Group Report, 35 FCC Rcd 3772 (5) (2020), accessed June 07, 2021, https://www.fcc.gov/document/fcc-approves-ligado-l-band-application-facilitate-5g-iot.

[32] Reardon, M., "Senators Pressure FCC to Reconsider approval of Ligado 5G," CNET, May 15, 2020, https://www.cnet.com/news/32-senators-pressure-fcc-to-reconsider-ligado-5g-approval.

[33] Eggerton, J., "NDAA Passage Puts Guard Rails on FCC's Ligado Approval," MultiChannel News, January 1, 2020, https://www.nexttv.com/news/ndaa-passage-puts-guard-rails-on-fccs-ligado-approval.; National Defense Authorization Act for Fiscal Year 2021.

[34] Manner, J., *Global: Telecommunications Market Access*. Norwood, MA: Artech House, 2002.

[35] Khatri, H., "Malaysia, Mobile Network Experience Report April 2020," Open Signal, accessed June 02, 2021, https://www.opensignal.com/reports/2020/04/malaysia/mobile-network-experience.

[36] Malaysia Wireless. "Malaysia Govt to build & manage its own RM-15bil 5G network with TM and Huawei," last modified Feb. 25, 2021, https://www.malaysianwireless.com/2021/02/malaysia-govt-build-own-5g-network-telekom-malaysia-huawei.

[37] Manner, J., "The Principle of Technology Neutrality in Telecommunications Regulations Must be Restored to meet Global Connectivity Goals," Via Satellite, accessed June 09, 2021, http://interactive.satellitetoday.com/via/december-2016/the-principle-of-technology-neutrality-in-telecommunications-regulations-must-be-restored-to-meet-global-connectivity-goals.; Manner, Jennifer, "Two Critical Factors to Ensure Connectivity to All," Space News, Sept. 24, 2019, https://spacenews.com/op-ed-two-critical-factors-to-ensure-connectivity-to-all/.

[38] Nokia, "Nokia selected by NASA to build first ever cellular network on the Moon," accessed June 09, 2021, https://www.nokia.com/about-us/news/releases/2020/10/19/nokia-selected-by-nasa-to-build-first-ever-cellular-network-on-the-moon.

[39] Roman, T. "6G vs 5G Network | Specs & Speed Comparison | What To Expect?," The Phone Talks, accessed June 09, 2021, https://www.thephonetalks.com/6g-vs-5g-network-specs-speed-comparison.

[40] Stuckman, P., "Europe Moves Towards 6G," Europa, March 4, 2021, https://ec.europa.eu/digital-single-market/en/blogposts/europe-moves-towards-6g.

2

The Evolving International Spectrum Regulatory Landscape

2.1 Overview of the ITU Spectrum Management Responsibilities

As we consider how to make adequate spectrum available for 5G and beyond through sharing and other mechanisms, it is important to understand the overarching international spectrum management regime at the ITU. The ITU is a treaty-based organization falling under the auspices of the United Nations system, with 193 Member States. The ITU has three major functions that include regulation and management of the global spectrum resource, setting telecommunications standards, and working with the developing world to help enable telecommunications.[1]

1. The other sectors, the Telecommunications (or "T-Sector"), which focuses on standards and the Development (or "D-Sector") which focuses on the developing world, all play a role in bringing 5G and future technologies to the world and are discussed throughout this book.

The Radiocommunication Sector (or ITU-R), is the body of the ITU with responsibility for spectrum management [1]. Specifically, the mission of the ITU-R is:

To ensure the rational, equitable, efficient, and economical use of the radio-frequency spectrum by all radiocommunication services, including those using satellite orbits; and to carry out studies and adopt recommendations on radiocommunications matters [1].

The specific responsibilities of the ITU-R include:

- The allocation of bands of the radiofrequency spectrum;
- The allotment of radio frequencies;
- The registration of radio frequency assignments and of any associated orbital position in the geostationary satellite orbit to avoid harmful interference between radio stations of different countries;
- To coordinate efforts to eliminate harmful interference between radio stations of different countries, and to improve the use made of radio-frequencies and the geostationary-satellite orbit for radiocommunication services [2].

The regulatory landscape for spectrum allocation and assignment is complex since spectrum is a shared resource belonging to no one [3] (but managed by national governments), understanding the ITU-R spectrum management process is critical to having a comprehensive view of how spectrum is made available domestically and internationally for different radio services. In this chapter, we will focus on the ITU-R and its spectrum governance responsibilities. In Chapter 3 we will discuss the domestic regulatory process and how domestic governments leverage the work at the ITU to determine their domestic radio frequency allocations. However, before delving too deeply into the ITU, it is important to understand the history of why the ITU-R was formed.

The radio spectrum resource was a largely unused resource at the turn of the 20th century and suffered from a lack of cohesive spectrum management. However, early in the 20th century, wireless technologies began to be used more often, including for safety of life services, such as maritime. In fact, the *Titanic* disaster in 1912 [4] exemplifies the spectrum issues that served as the driving force to have the ITU serve as the global spectrum manager. While there is a lack of clarity as to what happened that night with calls for help over the radio spectrum, the lack of interference-free communications between a nearby ship and the *Titanic* on the high seas resulted in the *Titanic's* distress signal not being received in a timely manner. This severely impacted the rescue efforts, resulting in more lives lost than necessary [5].

Following the outrage of the sinking of the *Titanic* with the rescue being hampered by the lack of adequate communications, another incident occurred. This was the lack of clear communications channels between Prince Henry of Prussia's ship and the United States when he was calling the White House[6]. This led to the international recognition that there needed to be a global solution for interference-free communications. This recognition resulted in the ITU taking over the primary role for regulation of the spectrum resource, and an increase in the number of domestic regulators being established with the role of regulating the domestic spectrum resource [6]. While much has changed in spectrum management since the early 20th century, the ITU remains the international bedrock of the international spectrum (and orbit) regulation and management. Today, the ITU is the global agency for spectrum allocations as well with this effort determining much of the future of spectrum use.

2.2 Overview of the International Table of Frequency Allocations and the WRC Process

To effectuate its spectrum management mission, the ITU-R manages the global International Table of Frequency Allocations, which is the global frequency band plan. The International Table of Frequency Allocations only change based on the output of treaty conferences, called world radiocommunications conferences (WRCs or Conferences) which are held when the ITU-R meets in plenary approximately every three to four years. The WRCs also take other associated actions regarding the governance of the spectrum resource including adopting resolutions and recommendations [6]. In between WRC, study groups meet to perform technical studies in advance of the conference.

The WRC process is often criticized for its infrequent meeting schedule since technology changes so quickly. In response, there was a movement in the early 1990s to have the WRC meet in two-year cycles. Unfortunately, it was found that this moved too quickly for the study cycle to be concluded. Then in the second half of the 1990s, the WRC resumed its longer three to four year meeting schedule. This delay in time between WRCs, means that the WRC process is not as agile as some would like. However, at the same time, since the WRC is primarily focused on allocations of frequency bands to radio services and not to applications (except for the identification of spectrum, as discussed below), some believe that having a faster process is unnecessary. The truth probably lies somewhere in-between the two, although many critics still complain correctly of the long lead times to effectuate spectrum changes within the ITU-R process. However, this time lag also helps to guard against ill-conceived changes to spectrum management that may have severe long-term impacts that are hard to reverse.

Each WRC agenda is finalized by the previous WRC, and includes a discrete number of agenda items that will be considered by the conference for adoption. For instance, for WRC-23, the conference agenda contains 11 items addressing issues as diverse as identifying additional spectrum for terrestrial international mobile telephony (IMT which encompasses mobile terrestrial wireless services such as 5G), allocating spectrum for narrowband mobile satellite service uses, and making spectrum available for global maritime distress and safety services [7]. Most recent conferences have addressed the need to identify spectrum for IMT. Due to the importance of the WRC genda, it often takes up a significant amount of time at the WRC, especially in the waning days of plenary meetings [8].

Although agenda items can be added to a WRC Agenda at the beginning of a conference, this is rare, because it requires the consent of the conference. This raises concerns as to whether there is enough study information to act on the proposed issue.[2] One of few times a conference agenda item was accomplished was at WRC-95. At that conference the United States proposed that the conference add for consideration an Agenda item to make 500 MHz of spectrum available for nongeostationary orbit (NGSO) satellite networks in the 28 GHz band. To ensure the successful addition of this proposed Agenda item, the United States had advanced some study work on this issue in the study cycle and actively sought political support. Because of these efforts, the United States did obtain sufficient support to have this issue considered at WRC-95. The United States was also successful in obtaining the allocation of spectrum it sought (although this took a second attempt, at WRC-97, to obtain the allocation for the full amount of spectrum) [9].

While at WRCs each member state has one vote, voting is a rare occurrence. Instead decisions are generally made by consensus arrived at through the negotiating process during (and around) the conference. This requires a robust discussion among the WRC delegates, often resulting in a careful balance in the outcome of agenda items (e.g., for instance, trades for support on one issue can be obtained by supporting another item that is important to the other party). Even as decisions on particular agenda items become more contentious due to their impact on incumbent services, there is a preference to avoid voting (it is still a rare occurrence). Consensus is seen as the best form of decision making to balance a variety of interests and, more importantly, to encourage member states to implement and follow WRC decisions [10].

Conferences are four weeks long and during the conference there are multiple meetings held each day, except for plenary, often simultaneously, at an attempt to come to consensus [11, 12]. For example, a typical day during the WRC can last 18 hours or more and have multiple simultaneous meetings on a host of issues

2. One challenge to establishing the next WRC's agenda is that until the current WRC is almost completed, it is near impossible to know which issues need to be carried over to the next WRC. Another short fall on the development of the next WRC's agenda falls on member states and the regional preparatory processes. In many cases, neither member states nor regional bodies give much consideration to future WRC Agendas because they ae focused on the pressing business of the current conference.

happening at the same time. This format is seen to benefit countries that have large delegations, such as Russia, China, Brazil, and the United States, because they have sufficient resources to cover multiple meetings simultaneously. However, countries that have smaller delegations, such as those in the Caribbean, can also maximize their participation, as discussed below, by working through regional organizations who represent their interests at the meetings. More importantly, even those with small delegations, understand the need to have the WRC break-down into small meetings to get the work of the conference done before reconvening in larger committees, and ultimately the plenary for final decision making.

As discussed, in the period between WRCs, there are intense study cycles where the technical preparation for the upcoming conference occurs. The study work to support the WRC Agenda items is assigned to individual ITU-R study groups, as detailed in Table 2.1, dedicated to specific technologies to perform the required technical studies to support each agenda item. The study groups are set up according to subject matter (e.g., Study Group 4 is focused on satellite issues, and Study Group 5 is focused on terrestrial uses). The studies are conducted by the member states and other ITU members. To obtain information on other radio services where the tudy group is not involved in a particular study, the study groups liaise among themselves. Joint study groups are also sometimes created, for example, when the work involves allocating or identifying spectrum to one radio service, but another radio service is likely to be severely impacted. This happened in anticipation of both WRC-15 and WRC-19, where spectrum was being examined for identification for IMT-2020.

Table 2.1
ITU-R Study Groups

Study Group	Subject
1	Spectrum management
3	Radiowave propagation
4	Satellite services
5	Terrestrial services
6	Broadcast services
7	Science services

In addition, while study groups cannot make decisions, their reports are critical inputs to WRCs and the conference preparatory meetings (CPMs), as study groups prepare draft CPM text for consideration at the CPM meeting held before an upcoming WRC [13]. However, unlike the WRC process where participation is member state driven, in the study group process, sector members can have greater sway and input. However, it is still exceedingly important to have member state support to be truly effective.

Study groups can also liaise with other bodies, such as standard bodies, to obtain and/or share relevant information. For instance, 3GPP, one of the leading standards bodies for 5G and beyond wireless standards, will often liaise with the ITU-R study groups to provide statuses or answer questions that can help inform the work of the ITU-R study process or visa-versa. With standards setting and the harmonization of spectrum for 5G and beyond becoming increasingly important, this open dialogue is increasingly important for the success of both organizations.

As shown in Table 2.2, the meetings of the study groups are held several times a year leading to each WRC culminating in the CPM, which is held approximately one year before the WRC and is where the CPM report is prepared. The CPM report provides a summary on the results of the study process, and provides views on the outcome of studies. The CPM meetings can be quite contentious as each interest is working to put the results that best favor them in the best possible light, into the final CPM report. In addition, the CPM report is critical because only a limited number of countries participate in the study meetings or even the CPM, largely because of resource issues. This is especially true of countries in the developing world. These countries often rely on the

Table 2.2
WRC Cycle

Meeting	Timing
CPM 1	Immediately following WRC
Study group meetings	Following CPM 1 until CPM 2
CPM 2	Approximately one year before WRC
WRC	3-4 years following previous WRC

CPM report as their primary basis for taking decisions at the WRC, especially on issues which they are less familiar or have less of a national interest in.

2.3 ITU-R Leadership and Participation

The ITU-R is led by the Director of the ITU Radiocommunications Bureau, an elected official who has a professional staff located at its headquarters in Geneva, Switzerland that performs the work of the Bureau. This includes supporting the meetings of the ITU-R and the work of the individual divisions, which can include processing international coordination requests, among other things.

Other organs of the ITU-R in addition to the WRC and ITU-R Director, include the Radio Regulations Board, an elected body, which handles a number of important activities such as managing disputes among members concerning interference, the ITU-R Secretariat, the CPM , the Radiocommunication Advisory Group, (whose work includes setting the strategy for the ITU-R, providing guidance for the Study Groups, and liaising with other organizations), and the Radiocommunication Assembly which sets and approves the studies to be carried out.

While countries are the only voting members of the ITU, private sector bodies, called sector members (e.g., telecommunications service providers, and equipment manufacturers), and regional operating agencies (ROAs), scientific and industrial organizations (SIOs), or financial and development institutions, can also participate, but do not have a vote. While sector members cannot vote on issues, many of them are quite active in the study period, where they participate, often influencing a number of studies based on their expertise. Sector members can also participate at more significant meetings, including the CPM and the WRC, but are not entitled to vote so their input carries less weight.

Sector members, SIOs, and ROAs can influence the WRC process through multiple methods. First, sector members can participate in ITU-R meetings from study groups to the CPM to the WRC, and more. In these bodies, sector members, SIOs, and ROAs can submit input in the meetings, as well as actively participate by making interventions on the floor. However, these entities are not entitled to a vote on the floor at WRCs, and hence their voices are not as important as those of a member state.

In addition, sector members, SIOs, and ROAs, as well as non-ITU members, can participate on one or more country delegations. Many ITU member states allow the private sector to participate on delegations to take advantage of the private sector's expertise in spectrum. In some cases, these delegates, if authorized by the member state, can speak at the country on behalf of that country at ITU meetings and often prepare and present member state inputs. Many sector members and private sector entities seek to participate on multiple member states delegations to help influence those countries make decisions that they support. However, such efforts are highly resource intensive so tend to be limited to either large multinational companies that have global spectrum interests or companies whose survival is dependent on an upcoming WRC decision.

ITU-R meetings are chaired by the members of the ITU. These chairs can be from a member state, a sector member, or a member of an RRO or SIO. The Chair serves at the highest level for a particular meeting, for example, the WRC or a study group. Under the chairman, there will be vice chairs, working group chairs, and potentially even lower groups, such as subworking groups or informal groups with chairs named. The goal of the subgroups is to reach consensus on a specific issue before moving it to the next level (though these issues can be raised to the next level even if consensus is not reached). ITU-R practice has shown that by addressing issues on a narrow basis, it is more likely an overall solution to a larger issue will be found.

Because of the importance of consensus in ITU-R decision-making, countries work hard ahead of WRCs to build alliances with other countries on issues of common concern where they share similar views. To increase their chances of success, countries meet regularly among one another, and in other ITU or non-ITU forums as they prepare for the WRC within their regional bodies to obtain support for their WRC positions.

There are also several regional bodies recognized by the ITU including[3]:

- *Asia-Pacific Telecommunity (APT),* Bangkok, Thailand;
- *African Telecommunications Union (ATU),* Nairobi, Kenya;
- *Caribbean Association of National Telecommunication Organizations (CANTO),* Port of Spain, Trinidad and Tobago;
- *European Conference of Postal and Telecommunications Administrations (CEPT),* Valletta, Malta;
- *Inter-American Telecommunication Commission (CITEL),* Washington D.C., United States;
- *Telecommunications Regional Technical Commission (COMTELCA),* Tegucigalpa M.D.C., Honduras;
- *Caribbean Telecommunications Union (CTU),* Port of Spain, Trinidad and Tobago;
- *League of Arab States (LAS),* Cairo, Egypt;
- *Regional Commonwealth in the Field of Communications (RCC),* Moscow, Russian Federation [14].

An important function of these regional groups is to bring consensus surrounding regional positions to the WRC and other ITU-R meetings. These groups can send representatives to the meetings who speak on the floor for the members of that group. Because they represent regional positions, their interventions carry a fair amount of weight and it's critical for advocates to get on board to be successful on a position for an agenda item. During WRCs, the regional groups meet quite frequently to ensure success in obtaining the desired WRC outcome. During recent WRCs they have been instrumental in negotiating the final outcome on critical issues.

3. These regional organizations have regular meetings, oftentimes several times a year, to have common positions on key issues before upcoming WRCs. Like the ITU, voting members are only governments, but they all allow for robust participation by private sector participants.

2.4 The Negotiations

Although consensus is the goal of decision making at the ITU-R, regardless of the body being discussed (from study groups to the WRC), there remains a clear recognition of the number of member states who support the outcome on a particular issue. Accordingly, in addition to garnering support from regional organizations, member states, and others (including the private sector), significant time and resources are spent reaching out to other countries to garner support for their positions on important individual agenda items.[4] This support can also be used at regional meetings to gain further support as well as to form alliances.

While the ITU-R process and WRCs are largely technically-focused, a fair amount of policy is advanced through positions and decisions taken, and politics are involved in the process, the deliberations, and the outcome. Political considerations may involve issues directly considered by the WRC as well as unrelated issues. For instance, at the start of every recent WRC the Arab world objects to Israel being accredited as a member state to the ITU and for participating at the conference. This issue is solved by a team of negotiators who spend the beginning of each conference working through a political solution. Therefore, before the WRC can progress, there is a mini-negotiation to resolve this issue so that Israel can participate, and the conference can proceed.

Political decisions often form the basis for many of the positions that member states take at the conference. For example, for leaders that want to bring 5G to their countries in the near term and want to advance vendors from their country, they may seek to include in WRC decisions the technical requirements that best support their domestic industry.

Oftentimes other nontelecommunications issues can be used as a basis for trade to support a particular issue if the member states involved so choose. And of course, countries may make swaps for support among agenda items to achieve their most important goals. The most intense trading occurs toward the end of the

4. For example, some of the countries who have the most at stake at an upcoming WRC will spend months ahead of a WRC traveling to meet with other countries to try and obtain their support for their positions. Other participants in the ITU-R, such as companies, will also do the same.

conference, with issues resolved through long discussions while balancing the different interests to arrive at a solution that will work for most, if not all interested countries.[5]

The outcome of WRCs is contained in the International Table of Frequency Allocations; allocations can be made on a primary or secondary basis. Primary services are entitled to protection from all services (other than coprimary services, unless otherwise provided for in the Radio Regulations), including secondary services and can cause interference with secondary services [15]. Secondary services are not entitled to protection from primary services, and cannot cause them interference [15]. It becomes more complicated when looking at the rights primary services have vis-à-vis one another, and for one or more secondary services. In these cases, countries engage in cross-border coordination to ensure that they do not cause interference into one another's lawful operations. These are often captured in bilateral agreements between the countries handled outside of the ITU-R process.

At the WRC agreed frequency allocations can be made on a regional global basis. There are three ITU regions: Region 1 which encompasses Europe, Africa, and the Middle East; Region 2 which covers the Americas; and Region 3, which encompasses the Asia-Pacific region [20]. The benefits of having allocations in multiple regions is for of the harmonization of the spectrum resource. For some services, like satellite services, there are significant efficiencies if spectrum is harmonized globally. Further, for example, for wireless services, global harmonization can result in increased economies of scale for equipment development and manufacture that can significantly reduce costs.

In some cases, countries have uses in frequency bands that are allocated on a global or regional basis at a conference for a different radio service they wish to keep in place for a domestic reason. In these cases, countries can take a country footnote at the conference. The inclusion of a country footnote to the Table of Allocations means that legally the country does not have to comply. However, a country footnote still means that the country taking the footnote cannot cause harmful interference into the operations

5. As discussed earlier, such political decisions may be subject to instructions of a country's leadership, and not left to the on-ground delegation.

of the country operating in a manner consistent with the allocation. This makes it easier for countries with large territories and few borders, like the United States, Australia, or Canada to successfully take these footnotes without negative repercussions. However, generally country footnotes are not looked at as they call into question the harmonization of that particular frequency band's use.

Since the WRC is a treaty-making conference, its final acts are treaty text. For some countries this means the text automatically becomes law. However, for others, such as the United States, implementation is necessary and requires notice and comment rulemakings. While some countries move quickly to implement the treaty into domestic law, others, due to procedural hurdles, lack of interest, or politics, may delay implementation. This means that private sector or government advocates for the allocation may have to appeal to the government to implement some of the WRC actions of importance to them; and even then, there may be no action. However, lack of quick action by countries to implement WRC actions can negatively impact the ability to deploy services in an efficient manner as the operators have to wait for domestic implementation to be complete before licenses are available. This is increasingly troubling as the need to access new spectrum becomes more critical to service providers.

2.5 The Role of WRC and the Identification of Spectrum for IMT Enabling 5G

As 5G is expected to be a network of networks, it is important to understand the different radio services which are likely to be a part of the 5G global network. These include, but are not limited to:

- The fixed service (FS) which is spectrum made available for point-to-point or point-to-multipoint fixed-terrestrial wireless services (e.g., microwave backhaul or high altitude platforms [HAPS]);
- The mobile service (MS) is spectrum made available for terrestrial wireless services that are mobile (e.g., smartphone use);

- The mobile satellite service (MSS) is spectrum made available for satellite service that is provided to mobile terminals (e.g., satellite phones like Iridium);
- The fixed satellite service (FSS) is spectrum made available for satellite service that is provided to fixed terminals (e.g., satellite broadband service to the phone (though this service has been expanded to include satellite service to Earth stations in motion on ships, planes, and the like).[6]

For 5G wireless services (terrestrial and nonterrestrial), the general preference is for harmonized spectrum accomplished by obtaining global or at least a regional allocation. Nonharmonized spectrum allocations can have a myriad of negative impacts including, for instance, not allowing companies to capture the economies of scale in the 5G equipment. Accordingly, in the most recent WRC processes, as 5G has become a focus, there has been tremendous pressure on ensuring harmonized global allocations to support the service and an increase in participation by equipment vendors and wireless service providers to advance harmonization.

Because the categories of radio services are broad and there is a desire for spectrum harmonization for more specific uses (i.e., applications), the WRC can identify spectrum for applications through a footnote to the Radio Frequency Table of Allocations. Identifications of spectrum is not legally binding (i.e., they are not treaty text like allocations are) but by including an identification of the band to a specific application, countries have a guidepost that this band is being developed globally for the particular application, such as IMT-2020 and beyond. While multiple identifications of spectrum have been made for different purposes, the identification of spectrum for the terrestrial component of IMT-2020 and beyond is for 5G use.

Table 2.3 shows all the bands that have been identified in recent years for IMT use in multiple regions. As we will discuss in future chapters, many of these bands are widely used for mobile

6. Other radio services, such as the Radio Navigation Satellite Service (more commonly referred to as GPS), may be part of the network, but will not be a primary form of communications, and are excluded from this analysis for that reason.

Table 2.3
IMT Terrestrial Component Spectrum Identified by WRCs

WRC	Frequency Bands (varies by Region)
1992	1885-2025 MHz and 2110-2200 MHz
2000	806-960 MHz, 1710-1850 MHz, 2500-2690 MHz
2007	450-470 MHz, 698-806/862 MHz, 2.3-2.4 GHz, and 3.4-3.6 GHz
2012	694-790 MHz
2015	470-694 MHz, 1427-1518 MHz, 3300-3400 MHz, 3400-3600 MHz, 3600-3700 MHz, 4800-4990 MHz
2019	24.25-27.5 GHz, 37-43.5 GHz, 45.5-47 GHz, 47.2-48.2 and 66-71 GHz

terrestrial wireless services today and are, or will be used, for 5G and beyond.

Although the concept of IMT identifications has been around since the early 1990s (made in response to previous generations of wireless services) the IMT-2020 radio interfaces for the 5G network were only completed in February 2021. They are contained in Recommendation M.2150, Detailed Specifications of the Radio Interfaces of IMT-2020 [16–18].

With 5G being deployed and estimated demand skyrocketing WRC-23 will consider identifying additional spectrum for IMT-2020 and beyond. The bands on the WRC-23 Agenda being considered for IMT include: 3600-3800 MHz and 3300-3400 MHz (Region 2), 3300-3400 MHz (amend footnote in Region 1), 7025-7125 MHz (globally), 6425-7025 MHz (Region 1), and 10.0-10.5 GHz (Region 2). In addition, WRC-23 is charged with an unusual Agenda Item, 9.1c which requires the study of IMT systems for fixed wireless broadband in the frequency bands allocated to the fixed services on primary basis. In addition, there is a preliminary agenda for WRC-27 that has also listed possible bands for IMT identification at that conference [19].

The identification of mobile service spectrum has become increasingly difficult, since the spectrum to be identified is shared with other radio services, and the identification for IMT-2020 in these bands will likely impact incumbent services [20]. For example, at WRC-19 the 24.25 GHz to the 27.5 GHz band was identified for the terrestrial component of IMT. Following the conference, a number of countries, including the United Kingdom, began consultations to make the band available for 5G mobile use with this frequency band being viewed as a key 5G mobile wireless frequency band. This may result in services that share these bands, such as FSS, either being relocated or extremely curtailed in their use of the frequency bands.

The WRC can also make other relevant identifications for 5G for other technologies. First, like for terrestrial wireless, there is an identification of the satellite component for IMT. This identification is for the Mobile Satellite Service (MSS). While MSS is seen as part of 5G, it is likely to be a smaller part of the 5G ecosystem than the FSS, in part, solely because of its narrowband capabilities. However, it may be very valuable for use cases such as the IoT or machine to machine (M2M) communications.

In addition, there is an identification for HAPS, which is expected to also form part of the 5G eco-system. Because HAPS operate at a fixed point, they are considered part of the fixed service, even if they serve mobile terminals. To date, HAPS has not been commercially successful, with even large companies, such as Google's Loon, going out of business.[7] However, as the need to deliver wide area 5G and beyond services increases, and technology development results in lower costs, we may see HAPS take off again.

Work has also begun on IMT for 2020 and beyond [21]. During 2021, it is expected that the ITU-R will define the schedule for future revisions of the Recommendation ITU-R M.2150, to accommodate future improvements to the standards, as well as the possibility of introducing new IMT-2020 radio interfaces. Looking even further ahead, ITU-R Working Party 5D has already started to examine future technology trends for IMT towards 2030 and beyond .This is also under discussion in the ITU's telecommunications sector, as discussed below [22]. This work will likely result in increased pressure on making additional spectrum available for 5G and beyond uses.

Accordingly, enabling the use of spectrum for 5G and beyond uses, as well as developing standards for the radio access network portion of wireless networks for 5G is an important part of the

7. Of course, IMT can be deployed even in bands not identified for IMT since it is part of the mobile service, depending on any technical requirements in the bands.

ITU-R's recent work. As demand increases, and as 6G is further defined, we expect this focus to continue.

2.6 Thoughts for the Future

The ITU-R will continue to remain an important body for managing the global spectrum resource. Spectrum harmonization will continue to be critical for almost all 5G and beyond technologies [23]. Accordingly, there will be increasing pressure to find ways to enable a more harmonized approach to frequency allocation, which may lead to additional pressure to move or limit the use of some frequency bands by incumbent services. For example, as clearing of bands of certain incumbents occurs in major markets, such as in the 3.5 GHz band for FSS in the United States, this may lead to additional pressure to clear the band for FSS to allow global use of these bands for 5G. However, all countries must continue to ensure that incumbent services that support 5G and beyond, as well as other critical incumbent services, such as those used for weather monitoring, can continue to operate on an interference-free basis. As the WRC continues to focus on making spectrum available for 5G, and ultimately 6G and beyond, it is important that the ITU and its member countries strike the appropriate balance in spectrum management to ensure that the spectrum resource is able to support 5G, 6G, and other important radio services.

References

[1] ITU, "Mission statement," accessed June 9, 2021, https://www.itu.int/en/ITU-R/information/Pages/mission-statement.aspx.

[2] UNOOSA, "Treaty on Principles Governing the Activities of States in the Exploration and Use of Outer Space, including the Moon and Other Celestial Bodies (Outer Space Treaty)," Jan. 27, 1967, https://www.unoosa.org/pdf/gares/ARES_21_2222E.pdf.

[3] For the definitions of allocation and assignment, see Riaz, Aamir, "National Table for Frequency Allocations," ITU, February 2016, https://www.itu.int/en/ITU-D/Regional-Presence/AsiaPacific/Documents/Events/2016/Feb-SMS4DC-Pacific/NTFA.pdf.

[4] Ryan PhD, Patrick Spaulding, The ITU and the Internet's Titanic Moment (July 13, 2012), *Stanford Technology Law Review*, Vol. 2012, No. 8, 2012, Available at SSRN: https://ssrn.com/abstract=2110509.

[5] Morrison, S., "The Radio Act of 1912 (1912)," The First Amendment Encyclopedia, https://mtsu.edu/first-amendment/article/1090/radio-act-of-1912.

[6] Struzak, R., "Introduction to the International Radio Regulations," https://www.osti.gov/etdeweb/servlets/purl/20945052.

[7] ITU, "World Radiocommunication Conferences (WRC)," https://www.itu.int/en/ITU-R/conferences/wrc/Pages/default.aspx.

[8] ITU, "RESOLUTION 810 (WRC-15) Preliminary agenda for the 2023 World Radiocommunication Conference," https://www.itu.int/dms_pub/itu-r/oth/0c/0a/R0C0A00000C0026PDFE.pdf.

[9] Manner, J., *Spectrums Wars: The Policy and Technology Debate*, Boston: Artech House, 2003, 186 pages.

[10] ITU, "FINAL ACTS OF WRC 1995," http://handle.i tu.int/11.1002/pub/800c948f-en,; Struzak, R., "Introduction to the International Radio Regulations," https://www.osti.gov/etdeweb/servlets/purl/20945052.

[11] Manner, J., *Spectrums Wars: The Policy and Technology Debate*, Boston: Artech House, 2003, 186 pages.

[12] Abernathy, K. Q., "Why the World Radiocommunication Conference Continues to Be Relevant Today," Federal Communications Law Journal 56, No. 2 (March 2004): 287–298.

[13] See ITU, "Conference Preparatory Meeting (CPM)," https://www.itu.int/en/ITU-R/study-groups/rcpm/Pages/cpm-19.aspx.

[14] ITU, "Regional Telecommunication Organizations," https://www.itu.int/en/council/Pages/rto.aspx.

[15] Struzak, R., "Introduction to the International Radio Regulations," https://www.osti.gov/etdeweb/servlets/purl/20945052.

[16] TA1LSX, "See ITU Regions and Map," https://www.ta1lsx.com/itu-regions-and-map.

[17] ITU, "M.2150: Detailed specifications of the terrestrial radio interfaces of International Mobile Telecommunications-2020 (IMT-2020)," https://www.itu.int/rec/R-REC-M.2150/en.

[18] ITU, "IMT -2020 Background," https://www.itu.int/en/ITU-R/study-groups/rsg5/rwp5d/imt-2020/Documents/060R1e.pdf.

[19] ITU, M.2083 : IMT Vision, "Framework and overall objectives of the future development of IMT for 2020 and beyond," https://www.itu.int/rec/R-REC-M.2083.

[20] ITU, "ITU-R Preliminary Studies for WRC-27," https://www.itu.int/en/ITU-R/study-groups/rcpm/Pages/wrc-27-preliminary-studies.aspx.

[21] Minter, A. "Google's Loon Failure Raised an Awkward Question," Bloomberg, February 6, 2021, https://www.bloomberg.com/opinion/articles/2021-02-06/google-s-loon-failure-raised-an-awkward-question.

[22] ITU, "ITU towards IMT for 2020 and beyond," https://www.itu.int/en/ITU-R/study-groups/rsg5/rwp5d/imt-2020/Pages/default.aspx.

[23] ITU News, "ITU Launches New Study on Network & Technologies for 2030 and Beyond," https://news.itu.int/itu-network-2030.

3

The Evolving Domestic Spectrum Regulatory Landscape

3.1 Overview

At the dawn of the telecommunications era in the 1800s, the first commercial telecommunications service was a wireline domestic telegraph. Shortly thereafter, the telegraph expanded internationally; later, it evolved into the wireline telephone [1]. It was not until the early twentieth century that radio-based telecommunications services began, which were quickly commercialized [2]. It was at this time that the events discussed in Chapter 2, such as the sinking of the *Titanic,* showed the need for international and domestic radio regulation/spectrum management [3].[1] Accordingly, at that time, governments began to regulate spectrum in earnest to support the broader international effort at the ITU to manage spectrum resources to ensure their countries' wireless communications did not interfere with one another.

While each domestic spectrum regulator is sovereign, there is a long-standing recognition of the need to coordinate spectrum regulation. This is recognized in the ITU-R Radio Regulations, which state that member states must "bear in mind that radio frequencies and the geostationary-satellite orbit are limited natural resources and that they must be used rationally, efficiently and economically, in conformity with the provisions of these Regulations, so that countries or groups of countries may have equitable access to both, taking into account the special needs of the developing countries and the geographical situation of particular countries." A recent survey of domestic spectrum managers' responsibilities and mandates around the world showed that these concepts are common goals of domestic regulation [4]. Accordingly, apart from focusing on ITU efforts at spectrum regulation, many domestic spectrum regulators also pay close attention to what other spectrum managers are doing. This is especially true when looking at neighboring countries, where inconsistent use may have a real impact, or when looking to other, more developed countries to assess the future of spectrum regulation.[2]

However, because of the increasing demand for access to more spectrum to support 5G and other services, we are beginning to see some changes to domestic spectrum regulation, notably the increasing need for harmonization, and the need to persuade incumbent users to abandon certain prized frequency bands. This chapter will explore the traditional structure of the domestic regulatory process, and some of the changes brought on by the need for greater spectrum access to sustain 5G and beyond.

---

1. For example, the United States began domestic spectrum regulation pursuant to the Radio Act of 1912.
2. For instance, many countries look to the United States and the United Kingdom, who are seen as spectrum innovators, as to how they are handling complex spectrum management issues. This has especially been the case with 5G as the United States has adopted some of the more innovative approaches to making terrestrial wireless spectrum available for 5G.

3.2 The Structure of Domestic Spectrum Managers

Every country has at least one government entity that is designated as the national spectrum manager. The structure of the domestic spectrum manager influences how new services are deployed. Domestic spectrum regulation/management may occur in a single agency, or the functions may be split across different agencies. For instance, the United States (Table 3.1) has a unique approach. The FCC manages the domestic Table of Frequency Allocations, commercial spectrum management, and commercial licensing, while the U.S. Department of Commerce's National Telecommunications and Information Administration (NTIA) manages government spectrum resources and licensing. When spectrum needs to be shared between the private sector and the government, these two agencies must coordinate. In addition, the NTIA hosts a body comprising all government spectrum users to guide its policies on spectrum management.[3]

Other countries, such as the United Kingdom, have a unitary agency approach, where one agency is responsible for allocating and determining the use of the spectrum for the private and public sectors, with appropriate input from each. In the United Kingdom, Ofcom assumes these responsibilities, in addition to issuing relevant spectrum licenses. A similar approach is taken in the United Arab Emirates, with the Telecommunications and Digital Government Regulatory Authority having sole responsibility for spectrum management.

Table 3.1
U.S. Government Agency Spectrum Responsibilities

Agency	Responsibilities*
FCC	Manages commercial spectrum (including public safety), manages the U.S. Table of Frequency Allocations, and licenses commercial users.
NTIA	Manages federal government spectrum, coordinates and represents government agencies in domestic spectrum process, and licenses government users.

*The U.

3. In addition, internationally, the U.S. Department of State has the policy lead, with NTIA and the FCC serving solely as technical advisors.

Differences in the structure of the spectrum regulator may impact how successful regulators are in allocating spectrum for 5G. For example, in countries with a unitary regulator responsible for commercial and government spectrum, there is no need for the extensive coordination that occurs with the United States' separate agency approach. Further complicating the U.S. process is that the FCC is an independent regulatory agency, while NTIA is an executive branch agency. However, when the executive branch policy is aligned with the FCC's goals, the process can run fairly smoothly, as it does currently, with the United States allocating more spectrum for 5G. However, when agency goals do not align, significant issues can arise that slow or stop the process determining allocation and use. This was the case recently when the FCC wanted to make spectrum available for 5G that was shared heavily with government users [5].

3.3 Managing the Domestic Table of Frequency Allocations

193 out of 195 of the world's countries are members of the ITU and, hence, committed under their treaty obligation to abide by the ITU's International Table of Frequency Allocations and the accompanying ITU Radio Regulations.[4] There are some rare exceptions to this, such as a country taking a footnote to the ITU International Table of Frequency Allocations. However, even in this situation, the country would need to ensure that its use that is nonconforming with the International Table of Frequency Allocations would not result in harmful interference to services in other countries operating in conformance with the Table. This is an important consideration for domestic spectrum managers as they consider what country-wide spectrum decisions to make. In most cases, the allocations in the International Table of Frequency Allocations are an important, if not the most important, consideration in making domestic spectrum allocations.

In some countries, the changes at WRCs that are made to the International Table of Frequency Allocations are self-effectuating upon that country's ratification of the relevant WRC Final Acts.

4. The two countries that are not included include the Holy See and Palestine.

Other countries, such as the United States, must complete public consultations before implementing WRC decisions, often involving the consideration of public policy and the adoption of technical and licensing rules to support the allocation. Public policy concerns include spectrum efficiency, the economic value of the service, how the public interest will be best served, and competition. Technical rules focus on in-band and adjacent band protection, as well as assignment of the spectrum to specific uses/applications (e.g., mobile telephony). For example, following the Final Acts of WRC-19, the FCC commenced the rulemaking process to implement the WRC-19 decision on allocating the 51.4–52.4 GHz band to GSO Fixed Satellite Service (FSS) gateway use [6], a band that the FCC previously proposed be used for terrestrial mobile wireless 5G services. Despite the new allocation and the lack of a country footnote by the United States, the 5G terrestrial interests have argued that the United States should continue to make this band available on a primary basis for 5G terrestrial services at the expense of use by the FSS.

Still, other countries may not move forward with updating their domestic Table of Frequency Allocations to be consistent with the WRC actions without a particular need being demonstrated or until a specific band/use is requested by a service provider. In these cases, an advocate, such as a terrestrial fixed-wireless or fixed-satellite service provider, may have to file a petition or other similar requests to have a relevant allocation made, or seek government support for use of a specific frequency band. This often means that there are delays in harmonizing spectrum globally and regionally that can impact the rollout of services.

In many frequency bands, there are allocations to multiple services on a primary or secondary basis. Where a frequency band is allocated to several services, the regulator may allocate the band to all the services consistent with ITU allocations; select which of those services may operate in its territory (and which allocation to make); or even decide to split the band into subbands, where each band is allocated to one or more services. Accordingly, in instances where the WRC adds a new allocation, a country may choose not to make that allocation and, instead, limit the band's domestic use to an existing allocation or nonconforming use, as long as interference with another country's lawful operations is not an issue. These are complex decisions made by the regulator that involve technical and policy considerations that are generally done through a public consultation or rulemaking process.

As previously discussed, identifications of spectrum in the International Table of Frequency Allocations are not treaty text. Some countries may still consider identifications as self-implementing, but for others, a consultation or other action may be required to implement the identification, if the government so chooses. Although identifications for IMT, for example, are not legally binding on domestic spectrum regulators, history has shown that tremendous pressure is often placed on countries to proceed with the identification domestically.[5] In fact, following WRC-19, we have already seen auctions around the world in spectrum that we identified for IMT-2020 and beyond [7].

Finally, as previously noted, since countries are sovereign, they do not have to comply with the International Table of Frequency Allocations and the accompanying ITU Radio Regulations as long as they do not cause harmful interference to the radio services of other countries operating according to the ITU's allocations and regulations. In these cases, countries will usually take a footnote to the International Table of Frequency Allocations, as illustrated in Table 3.2, but others may choose to simply make the spectrum available for a nonconforming use.

Table 3.2
Example of Country Footnote (5.317)

Region 3	
610–890 MHz	Fixed Mobile 5.317 Broadcasting

5.317 Additional allocation: in Region 2 (except Brazil, the United States, and Mexico), the frequency band 806-890 MHz is also allocated to the mobile-satellite service on a primary basis, subject to agreement obtained under No. 9.21. The use of this service is intended for operation within national boundaries. (WRC-15)

5. For example, the United States had made available spectrum in the 1.9 GHz band for its own personal communications services (PCS) as opposed to IMT-type services, to which the band was globally identified. This led to a complaint by the EU to the United States. But, because IMT is an identification, the United States was wholly within its rights.

Even with a country footnote in the International Table of Frequency Allocations, there may be political backlash from nearby countries for taking such action where policy or other conditions make the allocation or identification particularly important to the advocates. Countries may occasionally take actions, without taking country footnotes, that are not consistent with the International Table of Allocations purely for domestic reasons. This is a rare case, but if it does occur, as with the case of country footnotes, there is significant pressure on countries not to proceed and to ensure that harmful interference is not caused to the lawful operation of radio services in other countries. If harmful interference does occur to countries operating consistent with the Radio Regulations, the offending country is supposed to cut-off the offending communications. If this does not happen, the affected country may choose to bring this issue to the ITU Radio Regulations Board or pursue political and/or diplomatic solutions.

In addition to spectrum managers, other branches of government, if appropriately empowered, can legally change the domestic Table of Frequency Allocations, which can impact operations in neighboring countries. An example of this, regarding the United States 2012 Spectrum Act, is shown in Table 3.3. Under this law, the United States Congress (not the FCC, which is the U.S. commercial spectrum manager) by statute reallocated 20 MHz of the 700 MHz frequency band for flexible use. This flexible use reallocation created a complex band plan for public safety broadband communications and commercial terrestrial mobile wireless [8]. This type of action can have an impact beyond the country's own border. For instance, the allocations made by the U.S. Congress in the 700 MHz band required the FCC to coordinate with both Canada and Mexico to ensure that operations in these bands would be

Table 3.3
700 MHz Band Plan

Frequency Band	Use
750–768 MHz	Public safety broadband
768–775 MHz	Public safety narrowband
775–788 MHz	Commercial mobile use
788–798 MHz	Public safety broadband
798–806 MHz	Public safety narrowband

free from harmful interference in the United States and would not cause harmful interference in the neighboring countries' services. This became even more difficult to do in Mexico, where the Mexican government rejected the U.S. band plan in favor of a band plan that was developed in Asia and being used throughout Latin America [9].[6] In situations like this, countries will work on a case-by-case basis to resolve interference or may decide to enter into bilateral agreements in which they agree to an approach that will minimize the opportunity for harmful interference from one country's lawful operations to another's. This may include approaches such as geographic separation of coordination among operators to limit the potential for harmful interference on a cross-border basis.

3.4 Assigning Spectrum for Particular Uses

Once an allocation of spectrum is made in the domestic Table of Frequency Allocations, the spectrum manager must assign that spectrum for use, and adopt the required technical and licensing rules [10]. The technical rules must include allocation of the spectrum, determine which uses are assigned to the band, and how they are authorized to operate in a particular frequency band, including power limits, coordination with other users, and the like. These technical rules depend on a number of factors, including how the regulator is going to assign the spectrum (e.g., license or authorize its use), as well as which uses are in adjacent bands. Accordingly, licensing rules are generally adopted simultaneously with the technical requirements for operations.

Most countries take this action as part of a public consultation or rulemaking proceeding (hereinafter referred to as a *consultation*). These proceedings are generally initiated by the government, but some countries allow initiation at the request of a member of industry. In either case, the consultation process generally allows the government to release a proposed use for the relevant frequency band along with accompanying proposed technical rules for comment, as well as proposed rules for the licensing process (applying for a license, the type of licensing process that will

6. The U.S. did not have the same issue with Canada, who chose to adopt and implement the U.S. 700 MHz band plan.

be utilized, license revocation, term length, and enforcement).[7] In seeking comment, the government may also ask questions concerning the requester's ability to achieve certain government outcomes, such as solving the digital divide. In some countries, interested parties will meet with the government during this process (such as in Mexico and the United States), while in others (such as Australia), no such discussions are allowed. When interested parties are allowed to advocate for their position, they often make policy and technical arguments in support of their position. These parties may also seek support from other parties, including elected officials, vendors, and users of their services. Following the consultation period, the spectrum regulator will issue a formal decision and will be positioned to start the licensing process, consistent with national requirements.

In many countries, reconsideration of these decisions can be sought if parties disagree with the decision and have an appropriate basis for the disagreement. Reconsideration requests may be heard by the spectrum regulator, by another government agency, or by a court, depending on the national requirements. However, no matter the process, once a decision is made, it is often difficult to overturn.

3.5 Domestic Licensing of Radio Services

After spectrum is assigned, the government is then able to make the spectrum available for use through a licensing process or on an unlicensed basis. It is important to note that not all countries use the spectrum manager to license frequencies—instead, this may be done by a separate regulatory body or, in some cases, the ministry responsible for telecommunications. There are three general licensing approaches for spectrum: the traditional approach, the market-based approach, and the spectrum commons (or unlicensed) approach.

7. One important short-coming in this process, which is rarely discussed, is that the examination of the potential for harmful interference from making a frequency band available for a specific use, is limited to a two dimensional analysis. While this may be a reasonable approach for the near term, as nonterrestrial technologies become increasingly important it is vital to consider a three dimensional analysis to interference.

The traditional approach: In this approach, a regulator has total control, including choosing which entities are assigned spectrum and exactly how it is used. A common traditional assignment method is the beauty contest, where several companies outline their proposed service, and the regulator chooses the one with the greatest socioeconomic benefits.

However, other countries utilize a first come, first served approach. This approach is often used for licensing point-to-point microwave fixed services, such as for cellular backhaul, and for licensing communications satellites.[8] Norway, where the Starlink NGSO system is licensed, uses a first come, first served approach for satellite network licenses as does the United Kingdom and a number of other countries.[9]

With both traditional approaches, licensees generally pay a fee up front (usually dramatically less than in a spectrum auction) and an annual licensing fee.

The market-based approach relies on market forces to determine the distribution of spectrum. With the market-based approach, regulators treat spectrum like a private asset, typically selling licenses at auction and, in some countries, allowing resale or lease of these licenses by their owners so market forces can regulate the efficient use and distribution of spectrum. The recent auctions around the world of 5G spectrum are a good example of this process at work and have raised significant income for governments.[10]

The spectrum commons approach, also known as *unlicensed spectrum* or *license-exempt*, allows anyone to use a certain band without prior permission as long as the transmission technologies

---

8. One interesting trend of note is an effort by at least one mobile terrestrial operator to argue that satellite licenses that have been traditionally licensed under a first come, first served approach, should be subject to auctions in bands where they share with terrestrial services. This approach has not gained much traction, but is something to watch for as the debate over access to spectrum continues to develop.
9. The United States uses a first come, first served approach for some satellite networks, but has used a processing round approach recently for MSS and NGSO licenses where entities are seeking access to the same spectrum resource.
10. In many countries, spectrum auction revenue is used to help the government's treasury by being used for other purposes. Accordingly, it is not always clear that spectrum assignment decisions are not, at least in part, made partially based on the ability of the country to earn revenue for other purposes. In situations where this is the case, radio services that are not auctioned, such as unlicensed use, may be at a disadvantage.

adhere to certain rules (sometimes called *politeness protocols*) that manage interference. Examples include Wi-Fi and Bluetooth. There are generally no fees associated with the unlicensed frequency band use, but, unlike for licensed spectrum, there is generally no interference protection.

As previously noted, licensing rules are generally the result of consultations that set: 1) the process for applying for and receiving licenses; 2) the technical and other operational requirements; and 3) any other conditions, including the term of the license and enforcement provisions. Spectrum licenses may include both the transmit and receive portions of the radio transmission, though some countries do not license the receive portion of the communications directly, as it is not within their mandate due to the inability to cause interference.[11] In addition, many countries impose additional obligations, such as build-out requirements, on spectrum licensees to ensure that the licensees use the spectrum resource. Conditions may also include additional interference protections they have to afford other in-band or adjacent band users. Further, conditions may include public policy requirements, such as contributing to universal service funds or administrative reporting requirements.

Licensees under the market-based or traditional approaches receive some guarantees of protection from harmful interference, whereas whole, unlicensed spectrum users receive no guarantees against harmful interference. This means that unlicensed spectrum is more widely available and easier to access, use but will likely have a lower service quality. In addition, these users are not subject to the conditions that are imposed on licensed spectrum users. Unlicensed spectrum users also operate in a regime of *noninterference/nonprotection* to allocated services. They share frequencies with similar devices, all of which have generally equal rights—that is, with none of them having priority over the others (operation is likely lower than secondary). Without a limit to the number of devices operating simultaneously in an area, average used bandwidth cannot be guaranteed (meaning QoS levels cannot be guaranteed). This unlicensed approach to spectrum use has

11. This action has resulted in issues with poorly designed receivers, which create increased potential for harmful interference.

always existed, but has grown significantly with increased demand for Wi-Fi.

As spectrum managers seek to make additional spectrum available for 5G, there is growing tension between licensed and unlicensed use of these bands, as both uses require higher capacity to meet user demand while seeking access to the same frequency bands. This makes the decision of which licensing method to use for a particular band for 5G quite contentious. This was seen most recently with the 6 GHz band, where a number of countries faced this choice [11] and had to choose with how much of this frequency band would be made available for licensed or unlicensed wireless use.

3.6 The Role of Spectrum Harmonization and Its Importance in a 5G and Beyond World

An increasingly common theme in recent years for radio-based communications technologies is the need for spectrum harmonization. Spectrum harmonization means that spectrum allocation and use are harmonized regionally or globally. Harmonization has always been important for all technologies because when countries border each other spectrum use is most efficient when the radio resource is consistently used for each of the relevant frequency bands across all borders. For example, Switzerland is bordered by Austria, Liechtenstein, France, Italy, and Germany. Other countries such as the United States (with just two bordering countries, Canada, and Mexico) and Australia (with no adjacent countries to cause harmful interference) have to worry much less about spectrum harmonization, at least from an efficiency/interference standpoint.

In addition, there have always been several radio services where globally and regionally harmonized spectrum is critical. A good example is FSS and Mobile Satellite Service (MSS) both of which are services that provide regional and global satellite services because of their wide coverage areas. Therefore, to use FSS and MSS spectrum most efficiently, it is critical that these satellite systems be able to land traffic in as many countries as possible. Failure to do so means that the satellite system is not able to operate as robustly as possible and is not able to capture the economies of scale associated with providing service to countries within its coverage area. It is for this reason that FSS and MSS allocations are often the focus of great effort to ensure global or at least regional harmonization.

While all radio-based networks benefit from harmonization, there are some additional considerations for satellite systems. The satellite industry benefits from the ablility to use the same equipment with the same frequency bands across the globe. They can also host many frequency bands. While terrestrial systems are increasingly capable of using Radio Access Network (RAN) equipment with multiple frequency bands to adjust operations to the bands available in a given country, satellite systems are less able to quickly adjust operations. Because satellite systems are launched into space and stay aloft for years, the weight of the communications payload is an important factor. Adding frequency bands adds weight to these networks, which affects costs and, more importantly, technical viability. Accordingly, most satellite operators select frequency bands where a critical volume of spectrum is available for satellite use in their service area.

For other technologies, we began to see an increased need for regional and global harmonization as the world became more of a global marketplace. Over time, it became clear that regional and global marketplaces were key for telecommunications equipment and services. This became especially obvious with 3G wireless services, and has increased in importance in the terrestrial, mobile, and fixed wireless marketplaces. This, combined with the growing use of harmonized international standards for deploying global radio-based equipment and services, has increased global connectivity.

Other services, such as high-altitude platforms (HAPs), have also highlighted the need for the global and regional harmonization of spectrum. During WRC-2000, the HAP community recognized the importance of access to spectrum in providing their services in multiple countries using the same RAN and user equipment. This led them to seek global and regional identifications of spectrum for their use in the International Table of Frequency Allocations.

It is extremely unlikely that we would ever go back to a world where harmonization was not important for any wireless technology. However, it will take significant time for the current fragmentation in the International Table of Frequency Allocations and the domestic tables to change significantly. Of course, for new allocations and identifications for spectrum, we can expect to see increasing efforts at WRCs and to encourage regional harmonization.

In addition, harmonization efforts will likely take place outside of the WRC when efforts there fail. A good example of this is a recent attempt by Canada in the CITEL region to develop a band plan for the 28 GHz band for 5G mobile wireless services. This process has been controversial because the 28 GHz band was rejected at WRC-2015 for consideration for identification for IMT-2020, as it was heavily used for broadband satellite systems. Despite this, Canada, the United States, and several other CITEL members have made the 28 GHz band available for use by terrestrial 5G wireless.

Canada's proposal for CITEL is to create a 28 GHz band plan for terrestrial 5G wireless use. Adoption of this band plan regionally would entice CITEL member states to move away from use of the 28 GHz band for broadband satellite services. Canada and its allies see this as important to gain the benefits of regional harmonization for 5G terrestrial in the 28 GHz band that they failed to gain at the ITU. This is likely to be a contentious fight and currently, it is too early to determine the winner of this spectrum war. However, if this effort is successful, we will likely see similar efforts on the regional level to harmonize spectrum outside of the ITU when there is a failure to achieve such harmonization at the WRC.

The need for spectrum harmonization is further enhanced through the increased need for standards-based communications for 5G and beyond services. 3GPP, a voluntary standards body, is largely considered the most important standards body for 5G and beyond standards. As part of this process, 3GPP creates standards for the RAN for specific frequency bands where there are allocations that support such use.

Because of the increasingly intertwined nature of the standards and spectrum allocations process, we are starting to see a closer relationship between the standards-setting process and the spectrum management process. For example, a recent evolution in the standards-setting process, which has traditionally been driven by the private sector, is the increasing participation by government representatives. Additionally, in making spectrum assignment decisions for 5G, governments have been considering the actions taken by standards bodies in these bands during the consultation process. Accordingly, we can expect this trend to continue as the standards-setting process increases in importance in the 5G and beyond world.

3.7 Evolving Trends That Impact Licensing Decisions in a 5G and Beyond World

As domestic spectrum managers attempt to meet the growing demands for the 5G and beyond spectrum, we are seeing certain trends emerge in the licensing process. These include: 1) a return to use of the licensing process to obtain public policy goals; 2) the development of sharing regimes to enable use of spectrum by 5G by curtailing incumbent use; 3) incentivizing incumbents to relocate to allow 5G use; and 4) the relocation of government users to free up spectrum for commercial 5G use. These approaches are discussed further below.

3.7.1 Obtaining Public Policy Goals Through the Licensing Process

Throughout the regulatory process, regulators have imposed conditions on licensees to meet public interest goals. However, for a number of reasons during the 3G and 4G eras, there was a movement away from imposing such requirements. These include driving up auction prices (meaning more revenue for governments) and easing regulatory burdens on licensees. Now, with access to 5G being perceived as crucial to economic development, we are seeing a return to imposing conditions on 5G licensees that will help meet public policy goals. For example, as the need of governments to ensure 5G coverage in rural areas increases, countries are seeking to impose broader coverage requirements on mobile service licenses in rural areas. France's auction for the 3.4–3.8 GHz band included requirements that the winners launch services in at least 2 cities before the end of 2021; that each carrier deploy 3,000 sites by 2022, 8,000 sites by 2024 and 10,500 sites by 2025; and that 25% of the sites in the last 2 stages be deployed in rural areas [12].

3.7.2 Limiting Incumbent Use to Enable 5G Use

Another trend we are seeing is governments limitating the use of certain bands by incumbents so that 5G operators can make greater use of this spectrum. For instance, in the United States, the FCC opened the use of a frequency band (27.5–28.35 GHz) traditionally shared between the FSS and terrestrial fixed-wireless-to-mobile 5G services [13]. The FCC did not throw out the FSS use but imposed severe geographic limitations on their siting on terrestrial ground infrastructure (e.g., gateway Earth stations) and imposed a condition that for those limited uses, they would be subject to a first-in-time, first-in-right regime for future licensees in these bands. This change severely impacted the availability of this band for FSS, limiting operations to the most rural portions of the country, but has also given 5G licensees wide access to much needed spectrum [14].[12]

A similar approach has been used by the FCC in the CBRS regime, where the FCC has allowed continued use of the CBRS spectrum by incumbent FSS operators, but forbids growth. We can expect to see governments continue to limit use of incumbents whom they see as not using the spectrum resource efficiently in order to provide more access to 5G uses in the same frequency bands.

3.7.3 Incentivizing Relocation of Incumbents

To enable greater use of the spectrum resource by 5G another trend in domestic spectrum management is to incentivize operators to leave their frequency bands. It has always been challenging for spectrum managers to require licensees to make the most efficient use of the spectrum resource or to surrender access to

12. Interestingly, in 2021, the FCC Chair specifically questioned whether the FCC's focus on the millimeter wave bands made sense. In an interview in Axios, she stated that if we had to rely on the millimeter wave bands for 5G, the digital divide would increase.

spectrum they underuse, even when regulators impose strict use requirements. However, with their focus on freeing up additional spectrum for 5G use, governments have been creating incentives for incumbents to relocate to smaller portions of the spectrum resource.

The most notable example of this incentive-based spectrum management, as discussed in Chapter 7, is the FCC's incentive auction for the 700 MHz band, where the FCC encouraged broadcasters to voluntarily relocate by allowing them to share in auction proceeds, and reimbursing them for relocation. This approach was recently used by the FCC to encourage satellite operators to relocate from a portion of the 3.5 GHz band. Similar actions have been taken elsewhere, such as in Nigeria, where the government and NigComSat reached an agreement for NigComSat to vacate the 3.5–3.9 GHz band, with relocation costs being paid out of the 5G auction receipts [15]. We can expect to see this approach continue, as there is increasing demand for 5G and beyond spectrum use.

3.7.4 Relocation of Government Users to Free up Spectrum for 5G

As pressure mounts on spectrum managers and governments to create increased access to the spectrum resource for 5G, a number of governments are relocating government users from important frequency bands for 5G. This effort has been a priority due to the widespread belief that government users lack the incentive to use their spectrum efficiently because, in many cases, they do not pay for spectrum or have other market-based incentives to use the spectrum resource efficiently [16].[13] Due to the unique government applications of spectrum, such as for the military, it is more difficult to relocate government users than standard incumbents—sensitive systems may be at stake, and costs will need to be covered for relocation to other bands if services are to be continued.

However, the mounting pressure to make 5G available often provides the political support needed to make these difficult decisions. For example, in the United States, the Pentagon has agreed

13. It should be noted that some countries, like the United Kingdom, do require government users to pay for spectrum.

with the White House on a schedule for freeing up spectrum used by the military for use by commercial 5G users [17]. Similar efforts are occurring globally and may result in a more efficient use of the spectrum resource.

3.8 Conclusion

Domestic spectrum managers have always been pressured to make spectrum available for new and increasing uses of the spectrum resource. Today, due to the rise of 5G and beyond and the increasing reliance on spectrum-based radio services, regulators are under more pressure than ever. While the process for domestic spectrum management has not changed significantly, domestic spectrum managers are implementing new mechanisms to help advance greater use of the radio spectrum resource to make sure sufficient spectrum is available for 5G. This includes providing incentives for incumbents to relocate to free up spectrum for 5G use or otherwise limiting incumbent use of spectrum targeted for 5G use. To meet the anticipated demands for spectrum to support 5G and beyond, these actions are going to become increasingly important.

References

[1] Kunene, G., "A History of Telecommunications: How Telecoms Became Another Interface," Vonage, https://www.vonage.com/resources/articles/a-history-of-telecommunications-how-telecoms-became-just-another-interface.

[2] Struzak, R., "Introduction to the International Radio Regulations," https://www.osti.gov/etdeweb/servlets/purl/20945052.

[3] "The Radio Act of 1912 (1912)," The First Amendment Encyclopedia, https://mtsu.edu/first-amendment/article/1090/radio-act-of-1912.

[4] https://digitalregulation.org/the-role-of-the-regulator-on-national-spectrum-issues/

[5] FCC Approves Ligado L Band Application to Facilitate 5G &IoT, www.fcc.gov/fcc-approves-l-band-application-to-faciliate-5G-&-IOT.

[6] FCC, Adoption of World Radiocommunication Conference 2019, Allocation for Fixed-Satellite Service in the Earth-to-pace Direction in the 51.4–52.4 GHz Band, RM 11781.

[7] See 5G Spectrum Auctions, https://5gobservatory.eu/category/5G-spectrum-auctions/.

[8] Middle Class Tax Relief and Job Creation Act of 2012, Pub. L. No. 112-96, §§ 6101, 6102, https://www.congress.gov/112/plaws/publ96/PLAW-112publ96.pdf.

[9] Inzunza, A. L., "The Canadian Spectrum Summit 2017," FTI, http://www.ift.org.mx/sites/default/files/conocenos/pleno/comisiones/adriana-sofia-labardini-inzunza/oficio19informecalgarycanadaali.pdf.

[10] Riaz, A., "National Table for Frequency Allocations," ITU, February 2016, https://www.itu.int/en/ITU-D/Regional-Presence/AsiaPacific/Documents/Events/2016/Feb-SMS4DC-Pacific/NTFA.pdf.

[11] Hetting, C., "It's official: EU releases 480 MHz of 6 GHz spectrum to Wi-Fi – finally", Wi-Fi Now, www.wifiglobalnow.com/news-blog/it's-offiical-eu-releases-480-MHz-of-6-GHZ-spectrum-to-wi-fi-finally.

[12] Tomas, J. P., "France Completes 5G Spectrum Auction," RCN Wireless, www.rcn.com/20201002/5g/france-completes-5g-spectrum-auction.

[13] FCC, Spectrum Frontiers Report and Order and Further Notice of Proposed Rulemaking, FCC 16-89, www.fcc.gov/spectrum-frontiers-r&o-fnprm.

[14] Hart, K., "Unpack Future of 5G," Axios, Ina Fried and Kim Hart unpack future of 5G with Jessica Rosenworcel—Axios.

[15] "Nigerian Government and NigComSat Sign MOU on 5G Spectrum," Tech Africa, www.techafria.com/2021/05/07/nigerian-government-and-nigsatcom-sign-agreement-on-5g-spectrum.

[16] Ofcom, "Section 400 Spectrum Fees and Penalties Accounts," at 3.

[17] Neczyr, D., "White House, Pentagon agree to fastest transfer of spectrum for commercial 5G use," *Fed Scoop*, www.fedscoop.com/pentagon-spectrum-transfer-5G.

4

5G Technology Overview

This chapter provides a high-level overview of the technologies that will likely make up the 5G network[1] with a focus on some technology differentiators of the new generation of wireless services and planned spectrum usage.[2] This chapter sets a baseline for further discussion of how spectrum management regulators are meeting these often-competing spectrum requirement for components of the 5G ecosystem.

5G promises to be an important stepping-stone in the evolution of wireless technologies bringing higher speeds, lower latency, the ability to support a larger number of user devices (i.e., machine to machine and the Internet of Things), greater availability (anywhere, anytime access), and resiliency to wireless communications.

1. This chapter is limited to 5G technologies; Chapter 8 touches further on 6G technology.
2. We are not providing an overview of each technology except to focus on the changes that we are seeing as technologies evolve to be part of the 5G eco-system.

As with previous generations of wireless technologies terrestrial mobile wireless will be the primary technology for 5G with some of important differences. In addition to operating in the frequency bands below 5 GHz, used for earlier generations of mobile wireless technology, 5G technologies will need access also to mid- and high-band spectrum to meet its promise. In addition, the 5G ecosystem will comprise more than just terrestrial mobile wireless technologies. Other terrestrial and nonterrestrial technologies will complement the terrestrial wireless broadband infrastructure to bring 5G users the performance and coverage promised.

The inclusion of nonterrestrial technologies is already occurring as standard bodies, including 3GPP and the ITU-T, work to include technologies such as satellites and drones into the 5G standard setting process [1]. This is demonstrated by Release 17 of 3PP scheduled for release in March 2022 which will contain both terrestrial and nonterrestrial standards and will be further complimented with the ITU Working Party 4B's beginning to develop the standards for the satellite component of IMT 2020 and beyond.

While in some cases these technologies will complement the 5G mobile wireless platform by operating on a stand-alone basis, in other cases there will be a need for hybrid technologies that combine more than one technology, and are transparent to end users, bringing the benefits of both technologies to users (e.g., low latency and high capacity). In addition, it is likely that the overall network topology of 5G networks will be comprised of multiple technologies all working together to deliver the 5G experience to the user.

In addition, the terrestrial wireless portion of the 5G network will utilize a range of spectrum in the low-, mid-, and high-frequency bands to support the promised performance and coverage of the networks.

The need for access to additional spectrum in a range of frequency bands for the terrestrial wireless portion of the 5G network to meet performance requirements, combined with the 5G ecosystem relying on a range of terrestrial and nonterrestrial technologies, means that spectrum access and spectrum efficiencies are a must for success. Below we discuss the spectrum needs of the different technologies that will compromise the 5G eco-system.

4.1 Terrestrial Mobile Wireless Services

Terrestrial mobile services will remain the primary form of wireless technology to bring the promise of 5G to the world. 5G will build upon the previous wireless technologies, such as 3G and 4G, including operating in the low band frequencies and having standard cellular deployments. In addition, as we have seen with 3G and 4G, the 5G ecosystem will be heavily dependent on standards set in organizations like 3GPP. Because of 5G's high performance requirements, mobile wireless technology has had to evolve to meet requirements for speed, latency, capacity, and reliability.[3] As discussed below, this evolution has included the need to access additional spectrum in all portions of the radio spectrum resource (including the use of low, middle, and high band frequencies), the use of small cells, and the use of massive MIMO (Multiple Input Multiple Output) technology to increase spectrum efficiency.

4.1.1 Use of Low, Middle, and High Frequency Bands

Traditionally, mobile terrestrial wireless services have been reliant on low band spectrum (below 3 GHz) to ensure wide coverage and adequate capacity for their users. The relevant frequency bands that have been leveraged for previous generations of mobile wireless technologies include the 600 MHz, 700 MHz, 800 MHz, 900 MHz, 1.7 GHz, 1.9 GHz, 2.1 GHz, and 2.5 GHz. However, with its mandate for anytime, anywhere communications, and the increasing demand for higher speeds and larger capacity services, 5G mobile wireless networks will likely utilize more spectrum than ever before. Accordingly, as shown in Table 4.1, the mobile industry is turning not only to the low bands (below 3 GHz), but, but also to mid-band spectrum (3 GHz to below 24 GHz) and high band spectrum (above 24 GHz) to meet growing user demands for higher speeds and greater capacity services [2]. This mix of frequency bands also has the benefit of having different technical

3. There are also many non-spectrum developments, but these are beyond the scope of this book.

Table 4.1
The Many Technologies of 5G

Technology	Examples of Use Cases
Mobile terrestrial wireless	Connecting smartphones Smart Homes Smart Roads IoT M2M
Fixed terrestrial wireless	Broadband to the home Backhaul for mobile terrestrial wireless networks
High altitude platforms	Localized services emergency response
GSO and NGSO FSS networks	Broadband to the home IoT M2M Emergency response Backhaul for terrestrial wireless networks
GSO and NGSO networks	IoT M2M Emergency Response Use on Drones

characteristics so that operators can purposely design their networks using spectrum best fit for each purpose.

To understand how these frequency bands will work in the 5G ecosystem for terrestrial mobile wireless services, it is helpful to consider the use of these bands in a real-world scenario. For example, one mobile wireless tower operating in the low frequency bands can cover hundreds of square miles with 5G service that ranges in speed from 30 to 250 megabits per second (Mbps) [3]. A tower operating in mid-band spectrum covers a several-mile radius with 5G that currently ranges from 100 to 900 Mbps [4]. Finally, a tower operating in the high frequency bands covers a one-mile or lower radius while delivering roughly 1-3 Gbps speeds [5]. Over time, each of these tiers are expected to improve in performance [6].

While low band spectrum has traditionally been used for mobile terrestrial wireless networks and has known attributes such as good propagation characteristics), there are several pros and cons of relying on mid- and high- band spectrum for 5G delivery. Mid-band spectrum is likely to be an important component of

Table 4.2
Examples of Low-, Mid-, and High-Band Spectrum Targeted for 5G by the Mobile Wireless Industry

Frequency	Category
700, 800, 900 MHz	Low-bands
1800, 2100, 2600 MHz	Lower mid-bands
3.5 GHz, 3.6 GHz, 4.9 GHz	Mid-bands
12 GHz	Upper mid-bands
24, 26 and 28 GHz bands	Lower upper-bands
38, 40, 48 and 50 GHz	Upper upper-upper bands

metropolitan area mobile wireless networks due to its wide coverage area. Although, it does have a wide coverage area it is not as broad as low band and has poorer propagation characteristics, although it can penetrate through walls which is important for indoor or urban coverage [7]. On the other hand, high-band spectrum operations provide a higher network throughput, up to 20 Gbps but have a significantly reduced range of approximately 300 meters (depending on the exact band) and they have smaller cell sizes. This poorer propagation means the users need to be nearly in line-of-site, while the reduced range means an increase in the number of required antennas thus increasing the overall cost [8]. Accordingly, as 5G networks are designed, it is important they consider the specific use case that must be met in choosing which frequency bands to utilize. Accordingly, in cases where higher frequency bands are used, it would be best to deploy smaller cells, while in other cases, another design may be more appropriate.

In addition because of terrestrial mobile wireless' need for additional access to spectrum in a wide-range of frequency bands, retaining access and gaining additional access to spectrum for other technologies in the low-, medium-, and high-bands targeted by the 5G terrestrial mobile wireless industry, including fixed terrestrial wireless services, satellite services, and a host of others (e.g., radio astronomy and Earth exploration) is becoming increasingly difficult. This is especially the case as governments globally prioritize the needs of 5G mobile wireless usage. Accordingly, while there are important sharing issues for the mobile wireless industry among themselves, the most contentious spectrum wars often occur when the mobile wireless industry is seeking additional access to bands that support other radio services. A good example of this as previously discussed is the CBRS frequency bands where incumbent FSS was essentially frozen.

4.1.2 Small Cells

Traditionally, terrestrial mobile cellular systems have relied on cellular networks with base stations that transmit over several miles. However, because of the variety of user cases and the performance characteristics that 5G needs to meet, radio network designers have to consider other approaches to deliver on the capacity and speeds required in a host of use cases. These include meeting the requirements from smart cities where other use requirements must be met to stadiums that periodically host a significant number of people with large capacity needs.

To meet these needs, wireless providers in the 5G world are and will continue to rely on small cells. Small cells work similarly to conventional cell wireless services including with techniques like MIMO and use of a mmWave spectrum (millimeter wave spectrum). However, unlike conventional cells, small cells are generally lower in power and cover smaller geographic areas from only several feet to at most a kilometer or two. They are also designed to be less complex and less costly, to allow for faster, easier implementation then traditional cell sites, generally being mountable for indoor applications or outdoor. As discussed below, there are three primary types of small cells: 1) femto cells; 2) pico cells; and 3) micro cells [9].

Femto cells are small cells that help extend coverage for residential and enterprise applications primarily to offload congested primary networks, extend coverage, and increase building penetration for indoor use. The coverage area is about 10 to 45 meters (indoors), they operate at about a 100 milliwatts of power and support 8–16 users per site.

Pico cells are generally used for providing in-building coverage in a small area such as such as hospitals, schools, offices, and shopping complexes, as well as on airplanes. Their coverage area is generally a bit wider than femto cells of about 100 to 300 meters (indoor) and they operate at about 250 milliwatts and support about 32-64 users per site.

Microcells provide the widest coverage area of the small cell types providing a coverage area of about a kilometer to 2.5 kilometers as they operate at higher power of 2-5 watts. They can support about 200 simultaneous users, but are more expensive than femto or picocells. Their use is likely to be seen in unique spaces within smart cities or transportation hubs, hotels, or malls.

Small cell technology offers the following technology benefits that are critical for 5G and beyond:

- High speed;
- High flexibility;
- Massive connectivity;
- Low latency;
- High reliability;
- Low to medium cost as compared to standard base stations [10].

Today, with limited deployments, it is still too early to tell how widely deployed small cells will be, especially because of the cost of wide-spread deployment, though at least some are anticipating a very large market [11].

4.1.3 Massive MIMO

Another critical technology being developed to increase the spectral efficiency of 5G networks, is massive multiple input multiple output (mMIMO). mMIMO is a technology that uses multiple antennas configured in a two-dimensional phased array [12]. The mMIMO antenna system is attached to a base station that controls the transmission and reception of radio signals [13].

mMIMO systems are systems with up to several hundred antennas. Multiple antennas working together provide several advantages:

- Higher gain;
- Resistance to intentional jamming;
- More paths to the 5G client provide stronger signal strength;

- Transmissions can be focused in a beam at 5G clients, which is called beamforming and provides more power to the client with less interference;
- More parallel antennas can serve a larger number of users;
- Antenna arrays can identify 5G client physical locations;
- Arrays can track mobile clients and direct the transmission beam at the client, following the client movements and maintaining network connectivity.

mMIMO systems can handle large volumes of network throughput and support large numbers of client connections, which is a core performance requirement for 5G networks [14]. We can expect to see massive MIMO to become an increasingly important part of the 5G terrestrial mobile (and fixed) wireless network used as a tool to increase spectrum efficiency.

4.2 Terrestrial Fixed Wireless Technologies

Terrestrial fixed wireless services, licensed and unlicensed, are expected to become a critical part of 5G networks. The role they play is largely dependent on the frequency band in which they operate with some bands having propagation characteristics to cover long ranges, such as is required for back-haul, and some bands better to support in-home use, such as Wi-Fi.

4.2.1 Wi-Fi

As we discussed previously, Wi-Fi is an unlicensed technology that has become a critical part of the terrestrial wireless infrastructure today and will likely continue as part of the 5G infrastructure. Traditionally, Wi-Fi has supported short range, easy to set up self-deployments in home and office environments using license-exempt or unlicensed spectrum, and has also served as a way to off-load cellular traffic [15]. However, more recently we have seen Wi-Fi used for purposes requiring longer ranges and wider coverage, as well as mobility and operator services using licensed spectrum [16]. The Wi-Fi community is active in the standards process to ensure that Wi-Fi increases its capacity to support 5G communications performance requirements including high capacity, lower latency, and faster speeds. A good example of this is Wi-Fi 6, which is based on an updated IEEE standard.

Today Wi-Fi primarily operates in the low- to mid-band spectrum. However, to meet the performance and capacity demands of Wi-Fi users, access to additional spectrum is needed. The Wi-Fi community is often at odds with the licensed terrestrial mobile wireless community over spectrum use as in the case of the 6 GHz band. As Wi-Fi becomes increasingly important and continues to see demand for its services increase, these battles will likely be more contentious as both sides (licensed and unlicensed) seek access to the same frequency bands.

4.2.2 Fixed Terrestrial Wireless Services

There are two key types of uses that fixed terrestrial wireless services will play in the 5G world: 1) fixed wireless access, whereby the fixed wireless network will provide internet access; and 2) back-haul for cellular sites.

4.2.2.1 Fixed Wireless Access

Fixed Wireless Access (FWA) is an established means of providing internet access to homes using wireless mobile network technology rather than fixed lines. FWA is critical to providing internet service where there is limited choice in service to the home. FWA can also provide point-to-multi-point communications to multiple users, as is used in a municipal wireless network [17].

Traditionally FWA has not been able to provide download speeds or latency levels that can compete with a modern fiber broadband connection, but that is changing in the 5G world. FWA is using 5G network technology, such as beam-forming and high-frequency millimeter wave spectrum, to provide a considerable performance boost to wireless broadband services, including providing faster speeds than previously offered [18].

In terms of spectrum use, FWA use has been focused in the mid-band spectrum of the frequency bands. For example, to provide superior coverage and moderate throughput, the 3-4 GHz band range is appropriate for low density suburban or rural areas [19]. However, that is changing with the improvement in technologies operating in the high bands (e.g., 24 GHz and above). For example, the 24-30 GHz range is now being targeted for use for FWA to serve higher density suburban locales and cities at fixed locations. FWA is already a proven technology in lower frequency bands such as the 3.5 GHz band to provide 5G type broadband services to fixed locations in less densely populated areas.

4.2.2.2 Fixed Wireless for Backhaul

Another critical technology required for the success of 5G is fixed wireless backhaul for mobile wireless (including small cells) and FWA networks. Fixed wireless backhaul occurs by creating a microwave communication link between locations, this can be building to building or last mile connections in rural locations, or a network presence for connecting devices such as client devices, SCADA devices, and the like [20]. Licensed and unlicensed (or license exempt) microwave links have existed for years in the mid-band spectrum but with 5G they have increased and are now utilizing high-band or the millimeter wave bands [21]. With the need for greater coverage and availability, including small cells, the use of fixed wireless backhaul for 5G will increase and, as noted for fixed wireless access, regulators will have to work to enable access to a wide range of frequency bands with competing services, such as mobile wireless 5G.

4.3 Nonterrestrial Technologies

Satellite and other nonterrestrial technologies are also becoming an integral part of the 5G and beyond infrastructure. As discussed below, as new satellite technologies and HAPs become a reality, these nonterrestrial technologies will play a bigger role in the 5G network. This is especially likely as leading standards bodies, like 3GPP, are currently including nonterrestrial technologies in their standards. Below we look at both satellite and HAPs technologies and the likelihood of their importance in the 5G ecosystem.

4.3.1 Satellite Networks

Like terrestrial networks, there are a number of satellite uses in a number of frequency bands that will likely form an integral part of the 5G network of networks. These uses shown in Table 4.3 are discussed below.

4.3.1.1 Direct to Consumer Broadband Services

The satellite industry and its role in communications are at a critical point. Satellite communications hold significant promise for 5G with improving technology in terms of performance capabilities providing higher speeds and greater capacity by utilizing spot beam technology, which allows greater spectrum resource.

Both GSO and NGSO satellites can play a role in meeting direct to consumer 5G needs on a fixed basis to the home. Each technology has pros and cons. For instance, GSO networks have higher latency than NGSOs, but until NGSO constellations are sufficiently robust, GSO satellites provide higher capacity services. GSOs are also able to provide service to wide areas of the globe, but unlike NGSO systems, GSOs are not able to cover the far northern portions of the globe where fishing and other important economic activities occur. The satellite industry is actively working to improve the performance of both systems working within 3GPP and other standards bodies to ensure that satellite technologies are part of

Table 4.3
Satellite Frequency Bands

Band	Service Allocation	Use
L band (1.6 GHz band)	MSS	Mobile services including IoT and M2M
S band (2 GHz band)	MSS	Mobile services including IoT and M2M
C band (3.4-7.1 GHz)	FSS	Cable head ends, international traffic, back-haul
Ku band	FSS	Cable head ends, international traffic, backhaul, mobility
Appendix 30B (planned C and Ku)	FSS	Direct to home services
Appendix 30/30A	BSS	Cable head-ends, international trafffic
Ka band (20/30 GHz)	FSS	Direct to consumer broadband, mobility
40/50 GHz bands	FSS	Direct to consumer broadband, mobility

the 5G standards ecosystem. In addition, both NGSOs and GSOs operate in mid-band and high-band spectrum today, and both require increasing amounts of spectrum to meet growing 5G user needs.

While GSOs use developed technologies including standard satellite antennas that are cost-effective, NGSOs must operate using omnidirectional antennas on their user terminals, which currently are much more expensive. In addition, some question whether the large expense to deploy such massive systems that require frequent refreshing due to the average life of satellites being five years will make economic sense [22]. This is especially true if terrestrial technologies can cost-effectively extend their reach through fixed wireless services, or if other nonterrestrial systems, such as HAPs, are technically and economically validated.

The spectrum for this use has primarily been focused on mid- and high-band spectrum. As capacity needs have increased, the satellite industry has looked to higher frequency bands, as have the terrestrial industry. In looking at spectrum issues, it is important to note that spectrum sharing issues among NGSO and GSO systems and between NGSO systems are complex. In addition, spectrum sharing between satellite and terrestrial wireless services raises a host of complex issues, as we will discuss in future chapters.

4.3.1.2 Satellite Direct to Consumer Mobile Services Operating in Cellular Spectrum

As the need to connect with rural users increases in order to deliver on the promise of 5G and beyond communications, several NGSO satellite constellations are being proposed that will reuse, with permission of the licensed cellular operator, the mobile service spectrum licensed to mobile terrestrial operators on its satellite system. These NGSO constellations claim that there is no need to make major changes to consumer devices to incorporate this technology. Several terrestrial operators are planning to work with these operators to extend the reach of their 5G mobile networks to more rural consumers who are not being reached with their terrestrial network [23].

However, there are multiple spectrum issues that must be resolved prior to full use of this technology. Foremost is that the cellular spectrum being proposed is not allocated to the fixed or mobile satellite service, but instead to the mobile service (e.g., 600 MHz). This raises a host of interference and other technical issues (e.g., what is the appropriate power level to protect adjacent operators). While under the ITU Radio Regulations, countries are allowed to use spectrum on a nonconforming use to the International Table of Frequency Allocations, they are not allowed to cause interference into adjacent countries.

There are also domestic spectrum issues that must be addressed. For example, if the spectrum is auctioned for cellular use, as in the United States, in small geographic areas, how do you handle interference between licensees that would arise from using the terrestrial spectrum on a satellite without appropriate technical protection rules? In addition, there has been no work to address possible interference issues from the use of mobile spectrum on a satellite system with adjacent services. It is likely that spectrum managers will have to address these and similar issues before authorizing the use of mobile terrestrial spectrum on a NGSO satellite system.

Nonetheless, while this proposed satellite service is quite nascent, if the spectrum issues can be worked out, it can help meet the anytime, anywhere requirements of 5G for the mobile terrestrial wireless industry at a potentially reduced cost.

4.3.1.3 Satellite Backhaul

For many years, satellite operators have provided backhaul to cellular networks especially in rural and remote areas. However, despite the availability of this service, cellular networks still rely overwhelmingly on fiber or microwave technologies for backhaul. However, with 5G, this could change for several reasons.

First, because of the need to connect more rural and remote areas of the world to the 5G terrestrial network, and the planned deployment of additional satellite communications systems, the cost-effectiveness and quality of satellite communications to support backhaul needs have increased. Second, for the first time, standards bodies, such as 3GPP, are including satellite technologies in their standards for 5G. This makes interconnection much easier and less costly than when proprietary technologies have been used. Further, with the increasing competition in the satellite industry, there is likely to be increased capacity both in satellite systems operating in mid- and high-band spectrum to provide this service. Fourth, with the improvement in GSO satellite technology to launch increasingly high-capacity satellites, there should be sufficient capacity to support the capacity needs for terrestrial backhaul [24]. Accordingly, we can expect to see satellite systems increasingly called on to provide backhaul.

However, it is important to note the spectrum used by the satellite industry to provide backhaul includes the very spectrum that the terrestrial wireless industry has plans to utilize for its own systems—including the C, Ku, Ka, and even higher bands such as Q and V band. This raises spectrum access issues that need to be addressed by spectrum managers.

4.3.1.4 Mobile Satellite Services

The satellite industry is focused on two other key areas of mobility as part of the 5G ecosystem: 1) mobility services to planes, ships, railroads, and trucks; and 2) IoT and M2M services. In both cases, because of satellite's ubiquity, satellite is well-positioned to provide these services as part of the 5G eco-system.

First, the mobility services (also known as Earth stations in motion or ESIMs), will be provided primarily through the use of fixed satellites in the mid- to high-band frequencies using the same GSO and NGSO satellite networks supporting direct to consumer and backhaul services. As demands for access to these frequency bands increases by these and other services/technologies, ensuring adequate access to this spectrum is critical to make sure that transport systems such as airplanes and ships have access to 5G services.

With IoT and M2M, while fixed satellite services, it is more likely these will be provided by the mobile satellite service either through GSO or NGSO satellite systems operating below 3 GHz band. While both GSO and NGSO systems can offer quality services, for latency sensitive traffic and when polar coverage is required, NGSOs will be the preferred solution. However, greater capacity is provided via GSO satellite networks. For these systems, while there have been spectrum sharing issues associated with terrestrial wireless services, the spectrum access for these systems is more certain.

4.4 High Altitude Platforms and Similar Technologies

One developing area of promise is the use of HAPs to bring advanced mobile and fixed communications to users. HAPS systems can potentially be used to provide both 5G fixed broadband connectivity for end users, and transmission links between the mobile and core networks for backhauling traffic. Both types of HAPS applications can enable wireless broadband deployment in remote areas, including mountainous, coastal, and desert [25].

These easily deployable stations operating in the stratosphere (layer of the Earth's atmosphere starting at 20 kilometers) are high enough to provide service to a large area or to augment the capacity of other broadband service providers. HAPS have become more viable due to the evolution of technology through advances in solar panel efficiency, battery energy density, lightweight composite materials, autonomous avionics, and antennas. Recent test deployments delivering broadband Internet access using stations approximately 20 km above ground have demonstrated their ability to provide connectivity to remote or underserved communities [26].

From a spectrum perspective, like any mobile base station, the HAPS system needs spectrum access for backhaul, which is primary to satellites, as well as to connect to its users on the ground. HAPs have certain important attributes that make them ideally suited for being part of the 5G infrastructure. First, they are operating closer to the ground than even NGSO satellite systems, hence, bringing lower latency, which 5G requires for many use cases. Second, there is a fair amount of spectrum that has been identified for HAPs use internationally by the ITU which at least at the start should position them well to meet capacity and speed requirements. Third, while satellites cover wide areas, the HAPs system allows it to focus capacity on smaller areas then even NGSOs but are much wider than towers for either terrestrial mobile or fixed wireless services. This gives the operator greater control over where to deploy capacity (although today's spot beams on satellite networks have greatly improved this for satellite operators).

4.5 Conclusion

This brief overview of the different terrestrial and nonterrestrial technologies in 5G and their overall spectrum requirements helped set the stage for looking at spectrum management in a 5G world. With so many competing uses, the regulator's job is to develop frameworks to enable the most efficient use of the spectrum resource, and the greatest public interest benefits, an increasingly complex balance to achieve in a 5G world.

References

[1] See "Non-Terrestrial Networks; 3GPP's TNT," Jessy Cavazos, Keysight Blogs, accessed July 5, 2021 at www.blogs.keysight.com/blogs/ind.entry.html/2020/10/13/non-terrestrial_netw-pUxV.html.

[2] "The Mobile Economy 2021," GSMA, accessed June 28, 2021, https://www.gsma.com/mobileeconomy/wp-content/uploads/2021/06/GSMA_MobileEconomy2021.pdf.

[3] https://venturebeat.com/2019/12/11/5g-in-2020-how-devices-and-networks-will-evolve-over-the-next-year/.

[4] https://venturebeat.com/2019/12/10/the-definitive-guide-to-5g-low-mid-and-high-band-speeds/.

[5] Id.

[6] Id.

[7] T-Mobile, "Why Mid-Band Matters", accessed July 5, 2021 at www.t-mobile.com/business/resources/5g-mid-band-spectrum; "Identifying Key Characteristics of Bands for Commercial Deployment and Bands Subcommittee" CSMAC, NTIA accessed July 5, 2021 at www.ntia.doc.gov/files/ntia/key_characteristics_subcommittee_final _report_nov_17_2017.pdf.

[8] https://www.howtogeek.com/428337/not-all-5g-is-equal-millimeter-wave-low-band-and-mid-band-explained/.

[9] https://www.landmarkdividend.com/femtocells/ or https://www.waveform.com/pages/femtocell-and-microcell.

[10] Id.

[11] Small Cell 5G Network Market Size, Share - Global Forecast, 2026 (fortunebusinessinsights.com) (last visited May 9, 2021).

[12] Cite. General info https://www.mathworks.com/discovery/massive-mimo.html more specific to this part https://www.bluedanube.com/blog/2019/02/11/massive-mimo-practical-considerations-and-complexity-drivers/.

[13] Cite. Id.

[14] https://www.ncbi.nlm.nih.gov/pmc/articles/PMC7284607/.

[15] This year, 63% of mobile traffic will be offloaded to Wi-Fi. "Discover Wi-Fi Value of Wi-Fi," Wi-Fi Alliance, accessed June 28, 2021, https://www.wi-fi.org/discover-wi-fi/value-of-wi-fi.

[16] https://blogs.intel.com/technology/2021/03/wi-fi-and-5g-synergistic-technologies-for-emerging-applications/ this may also work for #19.

[17] https://www.academia.edu/2850931/Evolving_wireless_access_technologies_for_municipal_broadband or https://www.researchgate.net/publication/340525911_5G_Fixed_Wireless_Access.

[18] Cite. https://www.academia.edu/30543214/Fixed_wireless_access_ .

[19] Id.

[20] "Understanding Fixed Wireless Backhaul Configurations", Joe Wargo, Alpa Omega Wireless Blog, accessed July, 5, 2021 at www.aowireless.com/blog/bid/42303/Understanding-Fixed-Wireless-Backhaul-Configurations#.

[21] Although the use of higher frequency bands have poorer propagation characteristics and also require line of sight to communicate. Identifying Key Characteristics of Bands for Commercial Deployment and Bands Subcommittee" CSMAC, NTIA accessed July 5, 2021 at www.ntia.doc.gov/files/ntia/key_characteristics_subcommittee_final _report_nov_17_2017.pdf.

[22] See e.g., "Large LEO Constellations; Will it Be Different This Time?", Chris Dahnick, et. al., McKinsey & Company, accessed July 5, 2021 at www.mickinsey.com/industries/aerospace-and-defense/our-insight/large-leo-satellite-systems.

[23] www.as&tscience.com; www.lynxtechnologies.com.

[24] "Satellite Services: An Integral Part of the 5G Ecosystem", ESOA, accessed July 5, 2021 at www.esoa.net/reportsandstudies/lorem-ipsum-dolor-sit-amit-consectetur-adipisicing-7.

[25] https://www.itu.int/en/mediacentre/backgrounders/Pages/High-altitude-platform-systems.aspx; https://www.redalyc.org/journal/3094/309446752001/html/.

[26] Id.

5

Drivers of Change, Use Cases Driving 5G, Future Challenges, and the Race to Be First

5.1 Overview

To fully understand why governments and companies have focused on enabling the use of greater amounts of spectrum for 5G, it is critical to understand what is driving this change. Unlike the transition from 3G to 4G, where change was incremental, the transition from 4G to 5G is expected to be a significant evolution in technology, moving from connecting people to connecting everything, everywhere. This chapter discusses the drivers leading the changes that are supporting 5G roll-out and use, the specific use cases already being deployed or envisaged in a 5G world, the main adoption challenges and finally, a short discussion about the race to be first in 5G among many of the countries round the world. If 5G lives up to its hype and delivers on expected performance, it can bring dramatic changes to many areas of people's daily lives. People will see the performance changes in areas such as work, school, entertainment, and health and public safety. However, there are many challenges governments and operators need to address to enable the vision of 5G. Nonetheless, the perceived benefits of 5G are seen to be so great that many governments are in the race to be first in implementing 5G.

5.2 Drivers of Change

As an increasing number of advanced technologies and applications rely on wireless networks the performance of 5G is appealing. The 3G and 4G technologies relied on today are often viewed as insufficient to support the growing need for high-speed and reliable advanced communications. While users may appreciate the current 3G and 4G speeds and coverage they are finding that oftentimes the speed doesn't meet the demands of today's technologies, especially for uses like gaming and virtual and augmented reality. In addition, users are oftentimes out of coverage when they need connectivity, and for a growing number of applications, the latency is troubling. Also, there is not sufficient reliability or security of the network [1]. These concerns are compounded further by users with very specific needs. For example, in public safety or medical uses where today's 3G and 4G network performance is increasingly unacceptable to communicate and meet the user's mission, 5G is being seen as an important solution to meet evolving health care challenges [2]. Hence, 5G is becoming a reality at critical time to meet the growing needs of wireless end-users.

As 5G is deployed, the world is already seeing certain changes in the use of wireless technologies. This includes more screen time for some 5G users compared to 4G users because of the improved experience and greater reliance on 5G. These changes are leading to a reduced reliance on other telecommunications technologies that cannot offer the 5G experience. For example, there are some reports of reduced indoor Wi-Fi use by 5G users due to better and faster connectivity with 5G networks. We are also seeing reduced reliance on wireline technologies for broadband to the home as 5G networks often offer faster speeds than older technologies like cable and DSL. Changes in applications use are also occurring. For example, in some places there is an increase in gaming as 5G subscribers are spending a greater time on cloud gaming then non-5G subscribers do [3]. We are also seeing a greater reliance on broadband for work and school, largely driven by the COVID pandemic.

Therefore, this is a perfect time for 5G, often labeled a transformational user experience [4]. This is due to the communications improvements such as faster processing capabilities and the cutting-edge technologies developed to support 5G. In fact, some of the technologies driving the communications evolution to 5G are critical to support the growing use cases and enable the following telecommunications experiences:

- *Transformed visual experience.* The ability to provide rich visual experiences through 5G at scale can revolutionize the way content providers interact with users and how users interact with one another [5, 6]. There is a good deal of focus around 5G on the development of new applications including streaming videos, gaming, augmented reality (AR), virtual reality (VR), and mixed reality (MR). For example, a nurse can now use AR to identify a person's veins for inserting IVs, making the process more efficient and accurate [7]. However, such experiences cannot be fully realized using legacy wireless mobile networks which fail to provide required speeds and low latency. These types of visual experiences require the sort of high speed, high capacity, and low latency which terrestrial 5G should be able to deliver [8].
- *Multi-user experience.* 5G can enable multi-user applications such as military training simulations and remote control of robots in warehouses or other industrial environments. For example, today there are few industrial sites that do not rely on robots for a multitude of requirements.

  Multi-user experiences also bring wider control of assets than previous generations of technology because of the connectivity potential of 5G. 5G performance characteristics for the multi-user experience include wide-area coverage and low latency. The service enhancements promised by 5G will enable the creation of a seamless eco-system and network to support the multi-user applications, as opposed to having to utilize separate communications networks for different purposes. This connectivity creates cost and operation efficiencies for businesses and large-scale operations, which makes adoption more likely.

- *Hyper connectivity*. 5G can meet the growing demand for connectivity to support a wide variety of devices with a variety of speed, latency, and coverage requirements. This includes everything from cameras that support public safety, to sensors that measure temperature, humidity, and assess air quality in homes and businesses. Legacy 3G and 4G mobile networks do not currently provide this capability. 5G is being designed to have the ability to provide the required wide-area, real-time interconnectivity and required speeds to meet the relevant networks' needs. For example, cities are looking to using 5G to support traffic management that requires wide-coverage of all roads and low latency to respond to congestion and other traffic incidents, as well as high-reliability [9, 10].
- *Autonomous technologies*. Autonomous technologies are predicted to become an important part of the future, and are beginning to be visible on our roads today. Autonomous technologies can meet multiple needs including increased productivity from the growing use of robots in factories or stores, as well as self-driving cars, which can increase safety and decrease energy consumption [11]. The successfully operation of autonomous technologies requires the low latency that supports real-time applications as well as increased reliability [12]. The 5G network is being designed to support these types of requirements.

These are just a few examples of how 5G, if successful, can change the user experience. 5G can incorporate more communications technologies into the daily lives of people across the world—for work, school, health care, public safety, and so much more. If successful, 5G can deliver on its promise. It has the potential to push the use of the communications network further than ever before and enable use cases not supported by previous generations of wireless technologies.

Before examining use cases, it is important to understand the performance characteristics that 5G is promising; wide-coverage high-speed, low-latency, high-capacity, and ultra-reliable and secure communications. However, it should be noted that not all of these characteristics are required for 5G; it depends on your requirements. Accordingly, to be a part of 5G no performance characteristics must be met by a single technology. For example, an IoT network may require wide coverage, but only be carrying non-latency, narrowband data traffic, and can rely on a narrowband MSS satellite network. To support video gaming at an apartment in a city, the user requires a telecommunications network that delivers high-speed, high-capacity, low-latency service, but does not need to be highly secure or have wide coverage.Accordingly, while many of the use cases discussed below require many of the performance characteristics of 5G, not all do [13, 14].

5.3 Use Cases Driving 5G

The technological evolution in capabilities afforded by 5G telecommunications increases user demands and positions 5G networks to support different use cases which depend on promised 5G performance characteristics. While mobile smartphones are and will be an important part of 5G success and demand, the performance improvements, which can be dramatic (e.g., allowing the download of full HD movies in a matter of seconds) are usually only marginal over a 4G device, and are not driving the evolution to 5G [15]. Instead, the most impactful service improvement from 5G networks is going to be in areas outside of consumer handsets such as remote monitoring of manufacturing, 3D rendering, VR and AR, and with the increased speeds and lower latency in M2M and IoT communications to create an increasingly connected environment. Below we provide some of the more significant use cases that are driving the development of 5G.

5.3.1 Healthcare

5G promises many benefits for connected healthcare both at home and for doctors in hospitals, as well as health care practitioners in the most remote locations. This includes:

- *Faster medical image transmission.* Receiving medical test results on a quick basis can be crucial for successful patient outcomes. However, today's imaging machines typically generate enormous files which 3G and 4G networks cannot support. Utilizing 5G within hospitals and between health care providers, wherever they are located, can speed up data transfer and improve access and the quality of care. With the growing trend for hospital consolidation and limited access to health care in rural areas this technology allows physicians to share images and test results in real time regardless of their location increasing the quality of healthcare. 5G has the potential to solve this challenge.
- *Telemedicine:* Even without the pandemic, telemedicine and telehealth were expected to grow exponentially to meet user and medical professional demands. The pandemic has shown just how important this model is to the health of the country. For example, in the first quarter of 2020, after the COVID-19 pandemic hit globally, the United States saw a significant increase in telehealth visits compared to the same period in 2019.

  5G and the increased speed it brings, coupled with telehealth policy changes, can provide increased health care access in a number of ways [16]. First, these changes can further improve access to healthcare in rural and remote locations where access to medical staff is lacking (such as ships or offshore oil rigs). In addition, with the adoption of 5G medical offices and at the patient's home, the quality of telemedicine appointments can increase due to the faster data exchanges and better video streaming quality critical to observe the patient to determine the best course for medical treatment.

  In addition, there are many benefits to doctors having increased virtual contact with patients, including improvement in mental and physical health. The wide-coverage and high-quality video connection that 5G promises can significantly help by providing a better video connection among doctors and patients.

- *Emergency Response Service including Bioconnectivity.* Emergency medicine can also benefit significantly through widely available, high-quality 5G. To adequately care for patients during an emergency, it is important that responders can provide hospitals with critical data from the patient including data, high-quality images, and video, prior to their arrival. When required first responders can hold real-time video conferences with emergency room staff enabling communications between ambulance equipment and hospital telemedicine tools. This will improve hospital care once the patient arrives, increasing the patient's chance for survival.
- *Real-Time Bioconnectivity through Wearables.* At home monitoring of patients through different wearables and remote monitoring IoT devices can improve treatment plans and deliver preventive care. This is becoming more widely deployed, and can include devices for monitoring the vital signs of a patient to scheduling treatment sessions, thus avoiding unnecessary trips to the hospital. Further, at-home bioconnectivity devices can collect essential health data remotely for doctors' to monitor and review; further eliminating visits to the hospital. 5G promises the performance characteristics to meet this challenge.

5.3.2 Multi-User Experience

Multi-user experiences bring wider control of assets than previous generations of technology because of the connectivity potential of 5G. 5G performance characteristics include universal coverage and low latency. These service enhancements can create a seamless eco-system and network to support multi-user applications as opposed to having to utilize separate communications networks for different purposes. The following examples show how this application will operate:

- *Smart Military Bases.* Military bases are in essence small cities. They can and will use 5G in many of the same ways. In addition, they will be able to use 5G for personnel services, inter-base autonomous transit, and base security. In addition, military training and readiness can be increased with 5G to support war games and combat preparedness. Further, as with ports, a network of high-definition cameras, IoT devices, and sensors could provide real-time data on perimeter security, asset management, and mission-critical systems. A 5G-enabled smart base can also benefit military families by providing better educational resources for military spouses and children, enhanced telemedicine services, and more alternatives for families to visit remotely with deployed spouses.

- *Smart Ports.* 5G has the potential to enable the creation of a connected port ecosystem that supports automated cargo handling, can direct ship traffic, improve safety, and manage energy use. This will reduce human error and increase the efficiency and safety of operations. For example, a command center utilizing 5G can use AI to process data from camera feeds, sensors, and incoming ships to prioritize cargo offloading to make optimum use of space in container storage. This may lead to gains in efficiency and cost savings by having ships spend less time in ports and more time transporting goods.
- *Smart Factories.* 5G can also enable the creation of interconnected, efficient, and semiautomated smart factories. Using 5G, factory devices, including IoT devices, robots and human personnel should be able to exchange data to collaborate effectively over the 5G network.

5.3.3 Hyperconnectivity

One of the biggest promises of 5G technology providers is the commitment to hyperconnectivity. 5G hyperconnectivity is expected to impact everything from day to day needs to wearables.

5.3.3.1 Consumer and Everyday Needs

One of the main uses of 5G will be enhanced mobile broadband services. In addition, it is expected that wearable communications (only a recent possibility), and increased reliance on smart-home technology will be important consumer uses of the 5G network. These uses are discussed further below.

- *Enhanced Mobile Broadband.* As previously mentioned 5G will likely result in increased demand for faster and more capable smartphones because of the promise to bring better coverage and lower latency; two key demands of today's smartphone users [17]. Further, increased speed for video streaming is expected to be a growing demand [18]. Also, the reliability and security that 5G promises is a critical performance criteria as more users rely on their phones for everything including banking, health care, and public safety [19–21].
- *Wearable Communications Technologies*. Wearable communications technologies are increasing for both personal and business purposes. From Fitbit devices, devices for fun or business such as smart helmets, or glasses; wearables are becoming more present in our world. Wearables are also present on the manufacturing floors and in training facilities [22]. 5G should be able to enable wearables to communicate seamlessly and continuously because of the ability to support high-speed and low latency communications, hence increasing their utility. This will allow wearables to have access to accelerated speeds for compute/storage functions. This will increase the usability of wearables compared to that available with 3G and 4G [23].

5.3.4 Business and Industrial Needs

A significant driver of the future of 5G is business and industry. As discussed below, these needs range from the use of M2M/IoT, to wireless industrial control and more.

- *Massive M2M/IoT.* The ability to connect everything everywhere is gaining traction especially for business. M2M and IoT 5G is anticipated to connect all parts of a factory as well as be used by utilities to communicate between all parts of the network providing information back to central command locations. The anticipated performance characteristics of 5G, including wide indoor and outdoor coverage capabilities, as well as high-speeds, low latency, and security are key to the success of this market [24].

- *Wireless Industrial Control.* A growing number of industrial and manufacturing enterprises have already incorporated wireless control into their infrastructure to increase efficiency and reduce costs. This is accomplished through the integration of wireless networks to support sensor data and controller systems to improve internal processes. However, 3G and 4G capabilities have not been sufficient for this use to flourish and be deployed as widely as anticipated. 5G has the potential to enable wireless industrial control in areas where it was previously impossible due to limits of wireless networks providing the low latency required for smooth operations [25].

5.3.5 Smart Cities and Smart Infrastructure

As urban population density continues to grow rapidly, governments all over the world are looking into digital solutions to address common problems such as traffic congestion, better public transportation, and more effective delivery of public services [26]. 5G may help bring solutions to these problems. For example, 5G can bring increased connectivity to the sensors that form part of a smart city ecosystem, enable rapid data exchanges between them, bringing more technical solutions into existence, and reducing costs. Below are a few examples of where 5G will be of particular benefit to meet the goals of smart cities and smart infrastructure:

- *Smart Streetlights.* 5G can be used to form a connected system that manages the schedule of streetlights and provide notifications when there is a malfunction or other issue. A smart streetlight network can reduce costs for cities and municipalities [27].
- *Connected CCTV systems.* Cities have become increasingly reliant on CCTV systems for functions ranging from monitoring traffic conditions to public safety using cameras on buildings, light poles, and more [28]. Using 5G technologies to connect CCTV systems, municipalities on a real-time basis can address problems from crime to accidents, and quickly provide

information to the relevant public safety agency, improving response time as well the response. Further, this data can be used to obtain real-time traffic conditions to better reduce road congestion. Since the volume of data from a CCTV system can be extensive and time sensitive, there is a need for higher speeds for data transmission and lower latencies than 3G and 4G can provide.

- *Intelligent Parking.* While intelligent parking has been deployed using 3G and 4G, using 5G to support smart parking can increase the benefits obtained. For example, 5G can obtain information on parking space availability faster [28]. It can also help cities/towns improvethe monitoring and enforcing of parking violations, increasing revenue to government coffers [29].
- *Connected Traffic Lights.* 5G can also enable deployment of real-time traffic management systems, avoiding congestion and accidents. With 5G performance, there is the promise that municipalities can have a more granular level of control over the flow of traffic in response to specific demand levels, obtaining information on a real-time basis. The integration of 5G into municipal networks holds the promise for municipalities to respond to real-time conditions (such rush hour traffic or congestions in specific areas), improving road conditions for drivers. Government officials can also use this information to address other needs, such as giving priority to public transport and optimizing the overall traffic flow to reduce stop-start driving, reducing pollution.
- *Smart Highways.* We are already seeing 5G support smart highways (or roads) improving traffic congestion and road safety in some markets [29]. While still being fully developed, smart highways will likely rely on a combination of sensors along-side the road and communications among vehicles to apprise one another of changing road conditions including road surfaces, weather, and actions of other vehicles [30]. To provide the benefits foreseen with smart highways, the performance characteristics of 5G, including its high reliability, wide-coverage, and its low latency are critical.

- *Smart Electric Grid.* The global demand for electricity continues to grow dramatically, as does the need to control pollution, and reduce energy consumption. To manage the demand and meet goals for efficient, cleaner energy sources, new technological solutions such as smart grids and virtual power plants (VPP) have emerged. VPPs are connected entities aimed at optimizing the energy flow throughout the electric network. They have been slow to develop at least in part because 3G and 4G are insufficient to manage these networks appropriately [31]. However, 5G is foreseen as able to support the creation of VPPs because it has the speed and latency necessary to support the network [32].

5.3.6 Autonomous Technology

Autonomous technology is being developed and deployed today in our homes, our vehicles, and roadway infrastructure. As 5G is rolled out, it is likely to be better able to support these important uses.

- *Smart-Home.* To date the lack of reliable, wide-coverage service has delayed the development and use of smart home devices [33]. 5G may support newer features such as added security that relies on virtual and augmented reality. For 5G to be successfully used to support smart-homes, it is critical that it has the wide-area coverage that is promised.
- *Autonomous Vehicles.* There is a growing trend to integrate internet services into automotive applications in all types of vehicles (from cars, to trucks, to drones). In fact, today a number of vehicles rely on 3G and 4G cellular services to provide software upgrades, navigation, and emergency services. 5G is envisioned to be able to host new applications such as vehicle to vehicle communications and autonomous driving which, if successful, may significantly change the way people travel.

There are four major benefits to using 5G to support autonomous vehicles. The first is the potential for high-speed data transfers to improve data processing. The second is the extremely low latency that 5G is expected to bring. Expected 5G latency is close to 1 millisecond, while the human eye takes close to 400 milliseconds to blink.[1] This means that an autonomous vehicle using 5G to communicate with another similarly-equipped vehicle could perform 200 calculations between vehicles compared to the blink of an eye improving transportation safety.

The third benefit is wide coverage. For autonomous vehicles to be successful, coverage on all roadways (whether highways or the most rural country roads) will ne required. 5G promises such coverage through the use of low-, mid-, and high-band spectrum network deployments [34]. Finally, reliability and security are incorporated into 5G standards. This type of security is important for users to trust the technology to provide the safety and security expected on the roads, airways, and waterways.

5.4 The Main 5G Technology Adoption Challenges

Different industries and cohorts of users such as consumers, doctors, city managers, and public safety personnel have the potential to benefit if 5G delivers on its promise. Though 5G services are gradually being rolled out by wireless operators in a handful of countries, there is still a long road ahead to widespread usage in many countries, much less the world. Some of this is due to the long-lead time for the development and deployment of 5G terrestrial networks. However, there are other challenges that include:

- Access to Spectrum. All wireless services, terrestrial or nonterrestrial, require access to additional spectrum to support the increasing demand for 5G services. For example, the terrestrial wireless industry has made it clear they need access to significant bandwidth in the low-, medium-, and high-frequency bands. In some cases, this spectrum is still being allocated for 5G use internationally and domestically, and in others, the assignment of this spectrum is still on-going [35–37]. In addition, other terrestrial and nonterrestrial wireless technologies,

1. There are multiple planned spectrum auctions to get this spectrum out around the world, although a number of countries, especially in the developing world, are not at this point yet.

including fixed wireless, HAPS, and satellite services, are in search of additional spectrum which has yet to be made available and licensed for these bands [38]. Until adequate spectrum is allocated and licensed to users to support 5G, there is likely to be some delay in deployment in a number of countries.

- *Cost.* The deployment cost of 5G for terrestrial and nonterrestrial networks will be in the trillions of dollars [39]. Due to the overall deployment cost, operators must have a business case that makes sense. This means that operators are going to require knowing they have sufficient users willing to pay the costs of 5G services. Because of this, it is likely that areas of the world, especially in poorer, rural, and remote portions of countries, may not have access to 5G in the near term or will require government funding. Governments will continue to struggle with this problem, as they have for previous generations of wireless technology, with the likely result being long lead times for many developing countries to see 5G terrestrial deployment. However, these rural and less prosperous areas may still be able to have access to 5G by using nonterrestrial technologies, such as satellite.
- *Privacy.* 5G will simplify access to consumers personal and location data creating consumer privacy concerns. Many 5G users fear that businesses will resell, or otherwise transfer their data to third parties. To enable wide-spread deployment, it is likely that operators or, if they fail to do so, governments will need to proactively address these concerns to encourage greater use of the 5G network. There is some on-going work in this area. For instance, the European Union and the United States are both actively addressing the issues of privacy associated with 5G [39]. The outcome of this work, will likely help set the standards for the rest of the world. But until these standards are developed, and there is increased user confidence in privacy, 5G adoption may be slowed.
- *Security.* 5G has raised several security concerns even above those seen by 3G and 4G services. For instance, 5G networks will enable wider usage of the new cloud/virtualization technologies such as software-defined networking (SDN) and net-

work functions virtualization (NFV) [40, 41]. These technologies are highly flexible and programmable in nature, meaning they are more vulnerable to security threats [42]. Finding the right approach and/or solution to 5G security will be paramount for adoption especially by businesses, including tele-authentication and key agreements, and exhaustible protocol-based secondary authentication.

Further, with 5G involving M2M and IoT communications, perimeter cyber defense implementations will not be effective as it may be impossible to grant access exclusively to authorized devices. Therefore, this openness, means that the risk and potential severity of cyber-attacks grows exponentially. In addition, the important role of software in network virtualization will give equipment integrators wider and deeper access into network operations (i.e., data), therefore raising serious questions about trust and reliability. While governments and 5G operators have taken some baby steps to address the issue of network security, these efforts will have to increase to encourage 5G use and safeguard the critical data that will be transmitted using the 5G network. One way security is being addressed is through inclusion in standards, such as 3GPP. But there remains an open question whether such actions are sufficient.

- *Access to Facilities and Infrastructure.* Whatever technology is used to bring 5G to the end user will require infrastructure deployment. For example, for terrestrial wireless this can involve placing towers on public and private lands/buildings, the need for infrastructure to support wireless and wireline backhaul, and the deployment of repeaters in structures to provide high-quality, reliable indoor services. For nonterrestrial services this will require terrestrial access for gateway Earth stations, sufficient backhaul capabilities, the ability to place antennas for user terminals at users premises, as well as launching satellites into space. Access to these facilities will require governments to reduce administrative burdens and requirements. The good news is that a number of governments around the world have started working to reduce these burdens in anticipation of 5G deployment [43, 44]. This effort will need to continue if 5G deployment is going to be wide spread.

- *Standards.* To ensure that 5G fulfills its vision, further development of international standards must continue. As previously discussed, a number of bodies, including 3GPP and the ITU are actively developing standards for 5G [45]. For 5G to be successful, it is critical that the standards adopted are cohesive and coordinated to ensure the. Today a number of standards bodies, including 3GPP and the ITU regularly liase among themselves on items of shared interest. Seeing the ITU and 3GPP working on a coordinated basis to develop these standards is a positive reflection on the progress of 5G deployment as it should result in increased harmonization among the standards groups.

While there are many critical challenges to the successful deployment of 5G, the good news is, as discussed below, that governments and operators are heavily incentivized to find solutions.

5.5 The Race to Be First

With all of the new technologies that 5G can support and so many exciting use cases, it is unsurprising that countries and companies are actively seeking to be the first to reach 5G [46].[2] Over the past couple of years, being first or a leader in 5G has been an important stated goal. Countries like Japan, South Korea, the United States, and France along with leading global companies from Huawei, Nokia, ZTE, AT&T, Vodaphone are vying for leadership in this race. The race to 5G has very important economic ramifications from countries and companies.

While national pride is important in being first in 5G for countries, being a leader in 5G development and deployment has very real consequences. First, telecommunications is an enabling technology which can drive growth across a country's entire national

2. However, there are situations, such as with the Apple iPhone, where launching later can also result in success. For example, Apple was late to 3G and the iPhone was a success story.

economy. It is estimated today that 5G networks will add trillions of dollars in economic value to the global economy, unleashing vast technological change; whichever country leads or is an early adopter is likely to recognize significant economic value [47]. For instance, 4G brought with it such new technologies as Uber, Snapchat, Google Maps, and 5G promises much more. It has the capability to connect people and to connect everything, everywhere at high-speeds and lower latency with greater security and reliability than telecommunications networks have been capable to date.

Terrestrial and nonterrestrial service providers and network operators, such as Vodaphone, AT&T, Telefonica, Verizon or T-Mobile, OneWeb, SpaceX, Hughes, and others, along with equipment vendors, including Nokia, Hughes, ZTE and Huawei, all want to be first in 5G deployment for economic reasons. First, there is a long history of success in telecommunications being largely dominated by first-mover advantage. By being an early entrant, as Japanese mobile operator NTT Docomo demonstrated with 3G, a company and their home country can achieve great economic value, reducing chances forsuccess by competitors at home and abroad. Therefore, these actors (countries and companies) are incentivized to consolidate their leadership position early to maximize their returns. We are already seeing this in the fight over who will be first to win the 5G race—China, the United States, Europe, South Korea, or Japan.

Additionally, telecommunications networks remain a strategic asset from a competitive and a national security perspective. The speed, quality, and quantity of data is essential to compete in the current digital world. In addition, the data mining of 5G only makes it more attractive with the possibility of untold financial value along with additional available data for a country's national security purposes. This ability to capture vast amounts of data from 5G makes being first and being successful even more important. Along with being early market participants, market leaders will be increasingly influential in the standards setting process.

For these and other reasons, governments and companies are pushing through the challenges addressed in this chapter, including overcoming barriers to access to additional spectrum needed to be first and successful in the 5G marketplace. One thing apparent from the start for 5G, as it was for 4G and 3G, is that the market leaders for 5G are bound to be the ones to benefit the most economically, and otherwise. It is for this reason, as we will see in Chapters 7 and 8, that countries are developing innovative ways to allocate and assign the spectrum necessary to ensure that 5G will be a reality in the near term.

References

[1] "Why We Need 5G Technology and What It Means for Society," *Government Technology*, https://www.govtech.com/sponsored/why-we-need-5g-technology-and-what-it-means-for-society.

[2] Fisher, T., "How Are 4G and 5G Different," *Life Wire*, www.lifewire.com/5g-versus-4g-4156322.

[3] Krofli, B., "This is How 5G Is Going to Take Mobile Gaming to the Next Level," *Tech Radar*, www.techradar.com/news/how-5g-is-going-to-take-mobile-gaming-to-the-next-level.

[4] Logli, N., "5G Transforming User Experiences on Devices," *Arm Community*, https://community.arm.com/developer/ip-products/processors/b/processors-ip-blog/posts/5g-transforming-user-experiences-on-devices.

[5] "UAE 5G users spend more time on cloud gaming, live video - and less time using wi-fi," *Gulf News Report*, June 07, 2021, https://gulfnews.com/business/markets/uae-5g-users-spend-more-time-on-cloud-gaming-live-video---and-less-time-using-wi-fi-1.1623062255326.

[6] Jie, Y., "How 5G Will Take Esports to a Whole New Level for Gamers and Fans," *The Wall Street Journal*, May 26, 2021, https://www.wsj.com/articles/5g-esports-gaming-11621870177.

[7] William, R., "31 Augmented Reality Examples in the Real World," *ARVR-TIPS*, March 17, 2020, https://arvrtips.com/augmented-reality-examples/.

[8] "How 5G Will Improve Virtual and Augmented Reality," *Verizon*, https://www.verizon.com/about/our-company/5g/how-5g-will-improve-augmented-and-virtual-reality; "How 5G will enable virtual reality 'storyliving'," *The 5G Exchange*, https://www.the5gexchange.com/author.asp?section_id=743&doc_id=765920&.

[9] *AT&T Business Editorial Review Team*, "How 5G will Impact the Transportation Industry," https://www.business.att.com/learn/tech-advice/how-5g-will-impact-the-transportation-industry.html.

[10] Flynn, S., "How 5G Will Improve Hyperconnectivity in Smart Cities," *Tech Acute*, www.techaccute.com/how-5g-will-improve-hyperconnectivity-in-smart-cities.

[11] Terdiman, D., "The Network of the Future: How 5G Will Power Efficient Industrial Applications," *GE*, https://www.ge.com/news/reports/the-network-of-the-future-how-5g-will-power-efficient-industrial-applications.

[12] *NTIA*, "Automated Vehicles for Safety," https://www.nhtsa.gov/technology-innovation/automated-vehicles-safety.

[13] *AT&T Business Editorial Review Team*, "How 5G will Impact the Transportation Industry," https://www.business.att.com/learn/tech-advice/how-5g-will-impact-the-transportation-industry.html.

[14] O'Donnell, R., "What Does 5G Offer That 4G Doesn't," *USA Today*, https://www.usatoday.com/story/tech/columnist/2020/02/28/how-is-5-g-different-from-4-g/4892301002.

[15] Koonin, L. M., Hoots, B, Tsang, C. A., et al., "Trends in the Use of Telehealth During the Emergence of the COVID-19 Pandemic — United States," January–March 2020, *MMWR Morb Mortal Wkly Rep 2020*;69:1595–1599. DOI: http://dx.doi.org/10.15585/mmwr.mm6943a3external icon. https://www.cdc.gov/mmwr/volumes/69/wr/mm6943a3.htm.

[16] https://www.forbes.com/sites/tmobile/2020/07/06/5g-to-the-rescue-the-future-of-better-information-and-communications-for-first-responders/?sh=6725588474b6.

[17] "What is Enhanced Mobile Broadband," *Sierra Wireless*, https://www.sierrawireless.com/iot-blog/what-is-enhanced-mobile-broadband.

[18] *Ericcson*, "A Guide to 5G Security Network," www.ericsson.com/en/security/a-guide-to-5g-network-security.com

[19] https://vksapp.com/blog/industrial-wearables-changing-manufacturing.

[20] https://www.mbtmag.com/home/article/13251685/expanding-usecases-for-mobile-and-wearable-technology-in-manufacturing.

[21] https://www.ihrsa.org/improve-your-club/wearables-continue-to-transform-the-fitness-industry/ or https://www.glofox.com/blog/fitness-technology-is-transforming-the-industry/.

[22] *Verizon*, "5G and the Future of Wearables," www.verizon.com/about/news/5g-and-wearables.com.

[23] "5G and IoT Will Be the Leading Paradigm Switch in M2M," at www.ursalink.com/en/blog/5G-IoT.

[24] "A 5G Network for Every Factory," *Siemens*, www.new.siemens.com/global/en/company/stories/industry/industrial-5g-network-of-the-future.com.

[25] "Powering Smart Cities with 5G Technology," *Eastern Peak*, www.easternpeak.com/blog/powering-smart-cities-with-5g-technology.

[26] "Smart Street Lights for Brighter Savings and Opportunities," *Intel*, intel.com/content/dam/public/us/en/documents/solutions-smart-street-lights-for-brighter-savings-and-opportunties.com.

[27] www.megapixall.com/cctv-security-cameras-towns-cities.

[28] Berg, N., "The Future of Intelligent Parking," *Bloomberg*, www.bloomberg.com/news/articles/2021-03-22/the-future-of-intelligent-parking.

[29] "Why 5G Roads are the Future of Transport," *Highways Industry*, https://www.highwaysindustry.com/why-5g-smart-roads-are-the-future-of-transport.

[30] Girling, W., "Virtual Power Plants: Flexible Energy for the Digital Era," *Energy Digital*, www.energydigital.com/technology-and-ai/virtual-power-plants-flexible-energy-digital-era.

[31] https://www.gsma.com/futurenetworks/wiki/smart-grid-5g-network-slicing/.

[32] Allen, J., "How 5G Could Make Your Home Smarter," *Tech Radar*, www.techradar.com/in/news/how-5G-could-make-your-home-smarter.

[33] *Verizon*, "What is the Latency of 5G?" www.verizon.com/5G/5G-latency.

[34] Reimer, J., "Do Autonomous Cars Need 5G," *Here 360*, www.here360com/automous-cars.

[35] Clementi, A., "5G Technology Will Cost $2.7 trillion by the End of 2020-Report", *Mobile Marketing*, https://mobilemarketingmagazine.com/5g-technology-will-cost-telecoms-27tn-by-the-end-of-2020-.

[36] https://www.morganlewis.com/pubs/2020/06/securing-5g-networks-in-the-eu-high-risk-vendors-or-high-risk-legislation.

[37] https://iapp.org/news/a/5g-to-raise-privacy-challenges-and-opportunities/.

[38] https://www.gsma.com/publicpolicy/wp-content/uploads/2020/07/GSMA_5G_and_Data_Privacy_July_20.pdf.

[39] https://www.researchgate.net/profile/Ijaz-Ahmad-27/publication/322753634_Overview_of_5G_Security_Challenges_and_Solutions/links/5a75aec1aca2722e4dedf166/Overview-of-5G-Security-Challenges-and-Solutions.pdf?origin=publication_detail.

[41] www.3gpp.org.

[42] www.itu.int.

[43] Li, K., and Park, J., "Who was first to launch 5G? Depends who you ask," *Reuters*, https://www.reuters.com/article/us-telecoms-5g-idUSKCN1RH1V1.

[44] "The Top 5G Network Countries Most Likely to Launch 5G First," *SDX Central*, https://www.sdxcentral.com/5g/definitions/5g-network-countries/.

[45] Gourtsilidou, M., "Top 10 Countries – Leaders In The 5G Market, 2020," *CEO World,* 2021 https://ceoworld.biz/2020/06/16/top-10-countries-leaders-in-the-5g-market-2020/.

[46] Chan, A., "A Brief History of 3G Mobile Communications Technologies," *XOXZO*, 2021, https://blog.xoxzo.com/en/2018/08/10/history-of-3g.

[47] https://thehill.com/blogs/congress-blog/politics/523064-ensuring-more-americans-have-access-to-5g-technology.

6

Meeting the Growing Demand for Spectrum for 5G for All Technologies

6.1 Overview

As discussed in Chapter 5, having access to adequate spectrum to support 5G is critical for all radiocommunications technologies that are part of the 5G ecosystem to be successfully deployed. Since the 5G network will likely be formed with a number of terrestrial and nonterrestrial wireless communications technologies it is increasingly important for governments to enable access to adequate spectrum for all technologies to meet the projected 5G demands. However, the wide use of the spectrum resource today leaves little room for new and increased uses. Accordingly, many competing demands for access to the same frequency bands require more intensive and efficient use, from amateur radio, to 5G and beyond. To meet these requirements regulators must revisit traditional means of spectrum assignment and use, and explore new methods to make sufficient spectrum available [1, 2].

As a result, we are seeing increased efforts by governments, and spectrum regulators in particular, to find innovative ways to increase access to the spectrum resource. Also joining this effort are communications services providers, civil society, equipment manufacturers, users, and academia. As discussed in Chapter 7, this is leading to innovative solutions to enable greater access to the spectrum resource for 5G and beyond, including greater sharing among different uses.

Despite this effort, in the wake of 5G there continues to be strong competitive reasons for some interests to try to limit increased access to additional spectrum [3, 4]. These reasons include attempts to limit potential competitors from gaining access to new frequency bands, obtaining access to spectrum for purely speculative reasons, or even having licensees sitting on spectrum to realize its value on their company balance sheet, or to hold in reserve for a potential future use.

For these reasons enabling more efficient use of the spectrum resource to support new 5G uses is often difficult. However, there are a number of factors in favor of this goal. First, many governments see enabling 5G as critical to their country's economic success through increased access to advanced communications capabilities, including broadband to support education, business, and other critical applications like healthcare. Second, for access to mobile terrestrial spectrum, many governments implement auctions to best assign licenses. These auctions can bring in significant funds that the government can utilize to achieve other public policy goals [5]. The potential for revenue from these auctions can lead a country toward making spectrum available for this use.

Therefore, many governments around the globe are spending significant resources internationally and domestically to determine how to most efficiently utilize the radio spectrum resource to enable 5G and beyond. In this chapter we discuss some of the key approaches being used, and then discuss some examples from around the world in further detail. We conclude by raising the issue of a new spectrum management model called a three-dimensional model that takes into account the growing use of spectrum on the ground, in airspace, and in space, along with the impact this usage will have on sharing models and requirements to be considered going forward. It is imperative that such new spectrum models be put into practice to enable the greatest use of the spectrum resource for 5G and beyond services while guarding against the real risk of harmful interference.

6.2 The Changing World of Spectrum Sharing and Efficiency

The private sector and governments face growing pressure to consider ways to increase spectrum sharing to support 5G and beyond requirements. To aid these efforts there have been developments in some areas to increase spectrum sharing and the efficiency of the spectrum resource by current and future spectrum users. While many of these efforts have been around for years in practice or principle, the current race to 5G has increased their importance with resulting improvements, along with with new approaches being developed. For example, the concept of licensed or unlicensed spectrum has been around for many years, but its importance has increased with Wi-Fi becoming a technology relied upon on a daily basis. The section below looks carefully at spectrum models that have been around for years, assessing their effectiveness at meeting the increasing demand for 5G and beyond spectrum access. Chapter 7 compliments this review and covers examples of how these models are being used and/or built upon to make additional spectrum available for 5G. Some more innovative models are discussed as well.

6.2.1 Licensed Versus Unlicensed Spectrum

As discussed earlier, a continuing debate in the world of spectrum is between the spectrum property rights model (i.e., exclusive licensing), and the spectrum commons view of the world. The spectrum property rights model takes the position that radio spectrum can be treated like land, and that private ownership of spectrum is viable resulting in exclusive licenses for access to specific radio frequencies. This has been the primary licensing approach for spectrum over the past 100 or so years with licenses giving exclusive rights to specific spectrum bands.

The commons model, on the other hand, allows multiple users on a nonexclusive basis in specific frequencies. The commons approach has been used for years for some common uses, but its importance increased as more and more devices were connected. With 5Gs concept of connecting everything, the commons model is perhaps more important than ever to enable greater spectrum efficiency.

The proponents of the spectrum ownership model argue that the spectrum should be allocated to the prospective spectrum holders primarily through market forces (i.e., auctions) [6]. These spectrum holders (licensees) then have exclusive use of the spectrum portion they possess, without the potential of harmful interference from other parties. The licensees are then able to trade their spectrum rights in a secondary market [6]. This approach allows spectrum not being used by the exclusive licensee to be made available to third parties who would pay the primary rights holder for access to this portion of their licensed frequency band, in total or in specific geographic areas.

In the purest view of the spectrum property model, the use of spectrum would be flexible, in that the authorized or licensed party could use the spectrum portion for any purpose. Accordingly, the focus in this approach is on transferring ownership of the spectrum from the government to private parties.

This approach began to be implemented first by the FCC in the early 1990s, when it began to utilize spectrum auctions to assign terrestrial mobile and other spectrum licenses where there was more than one interested party (i.e., where the applicants were mutually incompatible) [7]. Since then, a significant number of other countries including Brazil, France, and Japan are successfully utilizing spectrum auctions [8].

In addition, the concept of secondary markets was implemented by the FCC for terrestrial wireless bands in the 2000s [9, 10]. Countries in Europe have also adopted and implemented this approach [11]. One of the key reasons for secondary markets was to encourage license holders not using their spectrum to have a financial incentive to allow third parties to make more efficient use of the spectrum [12].

However, the secondary markets approach has not been widely successful with less spectrum leasing occurring then policy makers expected. There are several possible reasons for this lack of success. These include the possibility that incumbent licensees are not interested in enabling new competitors in the marketplace, or the realized value received by the license holder for the lease does not justify the transaction. In some cases the licensees may be making access to spectrum available only in geographic areas that are least attractive.

The spectrum commons approach, which is based on the longstanding unlicensed use of the radio spectrum for a number of uses, is based on the view that as smart technologies evolve, communicating devices will become able to avoid interference through mutual cooperation and coexistence, and the spectrum will become less scarce [13]. The emergence of cognitive and software defined radio concepts, multiple antenna and multicarrier techniques, as well as ultra-wideband (UWB) technologies and mesh network topologies demonstrate this possibility. Many powerful advocates of this approach include key Silicon Valley companies, such as Microsoft and Intel. Under the commons model, communicating devices can share a specified spectrum band even without any one owner. Sharing of a common band of spectrum is done through the enforcement of technical restrictions and multiple access protocols [14]. The spectrum commons model is often analogized to a highway, which motorists treat as a common property and efficiently share as long as they abide by the traffic rules. However, for the spectrum commons approach to be successful, as in the highway model, there is still a need for a controller to ensure the parties have basic rules to follow [14].

The most successful example of a spectrum commons model of spectrum sharing is the use of unlicensed or license exempt frequency bands, such as the 2.4 GHz and the 5 GHz band, for Wi-Fi. Today, these bands are work horses that support Wi-Fi uses for broadband in the home and connect numerous devices [15]. In fact, many terrestrial mobile 5G devices are dependent on Wi-Fi for functions including: a user connected to a device via earbuds, a means to connect indoors to capture cost savings, or improved connectivity over the 5G mobile network [16].

Even with these two spectrum-use models in play today there continues to be a divide between the licensed (exclusive use) and unlicensed (commons) approach. Spectrum advocates have different views for use of the bands, including at the ITU, and in domestic regulatory proceedings. While almost all operators and governments agree that some split is appropriate, the disagreements often turn on the specific frequency bands, and how much spectrum to allocate to what use. In addition, these decisions cannot be made in a vacuum without looking at the impact of the planned band's use on other services. A good example is the decision to enable the 5925 MHz to 7125 MHz band (the 6 GHz band) for 5G mobile or unlicensed radio local area network (RLAN) use at the ITU and domestically. Since this is a band already used by other services, including satellite and fixed service among others, it is one of the concerns that have come up about how to protect this important incumbent use [17, 18]. In addition, the need for additional spectrum for each service must be reviewed, with both the RLAN and 5G terrestrial community making reasonable cases for their needs [19, 20]. There appears to be a demonstrated need for additional Wi-Fi spectrum; having additional harmonized spectrum for 5G coming out of WRC-23 would certainly help meet the demands of the 5G mobile industry for access to additional globally harmonized mid-band spectrum. Some countries are looking for a balanced solution and have split the spectrum between uses, while others are still deciding or have made the band available for a single use [21].

It is difficult to surmise how similar future battles between unlicensed and licensed uses that compliment and compete against each other might end up. One question may help drive solutions. We already see less Wi-Fi use with mobile 5G devices as indoor quality improves with the addition of more cell sites, and more capacity. Some believe that if this continues, there may be less need for access to Wi-Fi and other unlicensed services. However, the cost and benefits of Wi-Fi remain very attractive, so it is difficult to believe that new uses will not be developed that will call for additional spectrum. Accordingly, in the long run, the decision on making additional licensed and unlicensed spectrum in a particular frequency band available will continue to be a difficult balancing act for spectrum managers.

6.2.2 Dynamic Spectrum Access

Using dynamic spectrum access (DSA) to increase spectrum use and efficiency has been around as a concept for some time [22]. Simply, DSA is a technology that enables radios to share multiple frequency bands without interfering with legacy and other protected wireless systems. DSA improves spectrum utilization in three dimensions: 1) frequency; 2) location; and 3) time [22]. Using three dimensions, DSA enables a network opportunistically to use a wide range of frequencies at points in time and space, when and where they are authorized and available.

When a noncooperative user is detected on the same channel, a DSA-enabled device immediately moves to an unoccupied channel [22]. Because many frequencies are utilized sparsely in a fraction of locations, DSA enables two or more networks to share a given spectrum band [22]. DSA also enables a wireless service provider or user to deploy more than one application or service in a specific band, hence increasing spectrum efficiency [23].

DSA-enabled devices accomplish frequency sharing through a novel combination of RF, signal processing, networking, and detection technologies. These are coupled with DSA software algorithms to provide more robust communications capacity than is available through static spectrum access practices; hence, increasing spectrum efficiency and use [24]. Specifically, DSA has the sensing and dynamic reconfiguration capabilities that allow it to leverage the use of white spaces in the use of the frequency resource. This allows DSA systems to support opportunistic transmission without requiring access to extra spectrum bandwidth.

To build DSA systems, an efficient power communication processor is necessary to support the high-speed operation of this model on low and high power devices. For example, mobile devices are low power. Therefore, DSA requires satisfying two contradicting needs: high processing speed and low power consumption. In addition, some bands are more appropriate for use of DSA depending on the number of incumbent users to be accommodated, whether they are operating on a fixed or mobile basis, or if they have complex power budgets [25].

Today DSA is already being implemented by a handful of governments to enable greater spectrum use primarily in bands that are viewed as being underutilized by incumbents. A good example, discussed below, is the U.S. Federal Communications Commission's (FCC) Citizen's Band Radio Service (CBRS) approach. In this model, the FCC uses a DSA system to allow sharing between licensed and unlicensed mobile devices. Commercial communications fixed satellites and radars used by the U.S military are sharing space in the C band.

6.3 Public Versus Private Networks

In the 1990s and early 2000s there was a continuing push to do away with government-owned communications networks. Since these were legacy communications networks, they were oftentimes former monopolies. By privatizing them and liberalizing the marketplace, countries began to see more innovative services (including the wide-spread introduction and use of mobile terrestrial wireless and internet-based services) and lower costs [26].

With the opening of the market, there was little reason to look to government-owned networks to provide communications services. Generally, private operators facing competition provided better quality service, more innovative services, and lower prices than the former government entities did. However, in the past ten years there has been pressure to return to a public model, first using terrestrial technologies. Initially, it was limited to areas, such as Brazil, for the internet backbone [27].

Yet, other countries have now followed. Australia has deployed a national broadband network (NBN) to ensure that people across Australia had broadband no matter where they were located, which included a government-owned satellite broadband network [28]. The NBN, however, has faced economic and service quality issues throughout its existence [29].

Further deployments continue as we move toward to 5G. For example, deployment of government-sanctioned and sometimes municipally-run broadband networks have also become a reality throughout the world. For instance, across the United States it is estimated that at least 450 communities have municipal broadband networks [30]. However, there are a significant number of failed broadband networks as well, including in Chattanooga, Tennessee and Burlington, Vermont, among other locations [30]. The large failure rate of these networks raises questions as to how good an investment government operated networks are, at least on the municipal level.

In addition, during the Trump Administration, it was proposed to create a government-owned and operated 5G network for the country [31]. The outrage over this proposal was significant with criticisms including whether this was the proper role of government, and if such a significant investment was needed [32]. Ultimately, this proposal was tabled [33].

However, in countries like Malaysia and Myanmar, government-owned wireless networks are being implemented for 5G. These government-owned and operated networks have the rights to the 5G frequency bands, and the government is able to lease or allow other rights to use the network. Since these networks are a recent development, it is too early to see how efficient the spectrum use is. Based on past experience with government-owned and operated networks [34], it is unlikely that spectrum efficiency will be a goal or focus since they will not be subject to any meaningful competition [35].

Another recent trend is increasing government investment in private communications networks. A good example is the recent significant investment by some governments in commercial NGSO satellite mega constellations including the United Kingdom (OneWeb) and Canada (Telesat). Such investments made in these networks raises questions as to whether these networks will receive benefits from the investing governments, including access to the spectrum resource. These government-supported networks could hamper competition from other mega constellations that do not enjoy the same preferences, harming consumers through higher prices and less innovative services.

6.4 Hybrid Networks and Challenges

There are several different versions of hybrid models that may increase spectrum efficiency. These hybrid models include the concept of mobile satellite service with a terrestrial component; a NGSO satellite leases spectrum for terrestrial mobile operators, then reuses it on the satellite to bring service to the more rural areas. By combining different technologies, such as fixed wireless and satellite, into a single service the cost/efficiency is transparent to the user, allowing them access to the performance characteristics of both technologies.

6.4.1 MSS with a Terrestrial Component

A model long been expected to produce meaningful spectrum efficiency is the licensing of mobile satellite systems with a terrestrial component, known as MSS with an ancillary terrestrial or complimentary ground component (hereinafter MSS/ATC) in the same frequency bands. Today the United States, Canada, Mexico, and the European Union have all licensed these systems, within their countries. Australia and Brazil are also implementing such regimes, and New Zealand and Saudi Arabia are seriously considering them [36].

The concept of MSS/ATC systems was started in the United States, where they considered the limited number of MSS systems, and focused on two frequency bands: the L band (at about 1.6 GHz), and the S band (at about 2 GHz). The United States and several companies recognized that MSS was an expensive service with a limited rate of return, so incentives were needed to encourage the deployment of MSS systems. While sharing between MSS and mobile terrestrial wireless systems in the same band had been examined before coordinated use between two separate systems was unable to be resolved. Advocates of MSS/ATC argued that if you created a single license for the two services, the use of a single operator would allow for the use of both systems. The goal being MSS would be used in more rural areas, and the terrestrial portion would be utilized in more densely populated regions [37].[1]

1. These systems will, however, utilize MSS and other appropriate satellite spectrum for their gateway, TT&C, and intersatellite link requirements.

Because of the expense of the satellite, early licensing regulators decided that the spectrum would not be auctioned. And in those places where auctions have been allowed, like Mexico, auction amounts have been low.

Despite this vision, regulators felt that they had to create a model that would: 1) ensure the satellite spectrum was really being used by the satellite; and 2) limit terrestrial use in a way that did not make those operators true terrestrial operators, causing concern by terrestrial mobile service operators in countries who paid significant sums for their auctioned spectrum. This approach led to many conditions (some of which have been removed—like the EU's network integration requirement) which have limited the success of this model. It is likely for this reason, that MSS/ATC has a checkered deployment. For instance, in the United States, for 30 × 30 MHz of spectrum, the United States had 8 licensees when the S band was licensed in the mid-2000s. However, that soon whittled away to two licensees [38]. In part this was due to speculation, and the recognition that a successful system required more than the small amount of spectrum (about 4 MHz paired) awarded to each licensee with strict requirements for having an integrated satellite-terrestrial handset. However, even the increased spectrum was insufficient, and the two licensees both went bankrupt, with a MSS service licensee, DISH Network (formerly EchoStar), acquiring all 30 MHz. The FCC then eliminated the dual-mode device requirement, and gave other flexibilities to DISH [39]. Today, DISH has two MSS satellites over the United States the ICO and TerreStar satellite networks [40]. In addition, DISH is using this spectrum to build an Open Ran (ORAN) terrestrial system in the 2 GHz band, as well as additional bands it has acquired at auction.

The other MSS/ATC band in the United States is the L band. The L band was traditionally licensed to Ligado Networks (Ligado) and its predecessors going back to the original U.S. L band licensee, American Mobile Satellite Networks [41]. The Ligado licensing history is quite complex for MSS/ATC, but since its original MSS/ATC license in the 2000s, it asked for flexibility for its ATC component, which raised interference concerns with adjacent GPS bands [42, 43]. Finally, Ligado was licensed. Today, Ligado has 2 older MSS satellites operating along with a high capacity advanced MSS satellite, Skyterra One. However, Ligado has not yet begun the deployment of its ATC services.

Similar difficulties have arisen in the EU for its S band MSS/ATC licenses. Both its MSS/ATC licensees missed their original deployment dates, with Inmarsat, not launching a satellite in time, and Solaris Mobile (now EchoStar Mobile Limited [EML]) suffering an anomaly [44]. The EU extended the milestones for both. EML chose to launch a standalone satellite in the S band, while Inmarsat is utilizing a small payload on another operator's satellite. Inmarsat has deployed an air-to-ground system in its band space in the MSS and terrestrial component, while EML has pursued more traditional IoT network for its MSS system, and has not yet deployed terrestrially. Part of the delay for deployment of these systems is the uncertainty contained in the EU Directives creating the MSS/ATC regime, including requiring network integration. This was especially true with a lawsuit, brought by a competitor, who challenged certain aspects of the EU and member states interpretation of the directives in granting Inmarsat the rights to deploy an air-to-ground system. In 2021, the European Court of Justice found in the EU's favor (other courts have found the same across the member states), which may bring added certainty for more wide-spread deployment [44].

Both S band and L band advocates, however, have been active in 3GPP, ETSI, and other standard bodies to enable use of these bands for both MSS and ATC for 5G technologies [45]. For example, the S band CGC portion has been incorporated into the regime standard, and 3GPP is working to include the MSS S band into its Release 17.

With the checkered history of MSS/ATC, it is interesting to see whether other countries will adopt such a regime. Currently, Australia and Brazil have preliminary decisions that MSS/ATC will be part of their regulatory regime for the S band, and Saudi Arabia and New Zealand, are also considering it [46, 47]. As these countries develop their standard regimes, there are important spectrum management lessons to be learned. First, regulators must refrain from imposing unnecessary, overly restrictive network deployment requirements including requiring network integration and/or a single device for both MSS and ATC. When this has occurred in the past, the spectrum has laid fallow for a longer period of time. Second, it may make sense to provide additional flexibility for sharing between the MSS and ATC portions of the network to enable greater efficiency of spectrum utilization.

6.4.2 NGSO and Mobile Spectrum

A more recent approach under development is a NGSO MSS system. Instead of operating an MSS system, as the MSS/ATC systems do, the systems under development propose to use 5G bands, including the 600 and 700 MHz band as the frequency bands on their payloads [48]. However, the NGSO system operators under development, most notably by Lynx and AST & Science, do not plan to have the spectrum licensed to them. Instead, they are advocating for a secondary markets approach to gain access to the spectrum where they will work with the licensed terrestrial mobile service operator to operate their spectrum on their satellites [49]. The proponents of these systems argue that these operators can then work with the NGSO operators to offer service in the most rural areas. Licensed operators will insist the licensed terrestrial operator within the geographic area offer 5G using the satellite as opposed to having to pay the expense of a terrestrial roll-out [49].

There are many unanswered questions concerning these systems. First and foremost, is what sort of service will be available if offered over the NGSO system (i.e., will it be comparable to the terrestrial service) and what type of changes will consumer devices require. Depending on the answer to these questions, there may be concerns over the economic viability of such systems. But more importantly for purposes of our discussion is the potential for interference into other systems. Since these systems will be operating in non-MSS spectrum, on a nonconforming basis, under the ITU rules, they will have to operate on a noninterference basis under Article IV of the ITU Radio Regulations [50, 51].[2] How will operations of these systems be limited between the operations in a country where such use is authorized and a country where use is not authorized. Normally, under international rules this would

2. In addition, you will have many more moving connected devices like cars, trucks, ships, and airplanes.

require international coordination. However, in a case where the satellite use is not authorized, and the NGSO MSS systems are operating under Article 4 of the ITU Radio Regulations, requiring them to operate on a noninterference basis, the country impacted by interference can simply request the MSS NGSO system's transmissions to cease.

Similarly, in countries that do wish to authorize this use, but who don't have nationwide licenses, how will protection be afforded between these licensees? While the proposed systems state they plan to operate in the service area of a licensee with whom they have an agreement, this does not provide any protection for operators in adjacent service areas. To date, there has been no clear proposal to answer these questions, nor have technical rules been examined that might be required to protect against harmful interference. Since these are valuable frequency bands often awarded at auction, it is incumbent on spectrum managers to ensure these systems are not allowed to cause harmful interference into adjacent bands or geographic operators.

Finally, from a standards perspective, there remain open questions. For instance, to date, there has been no movement in the standard bodies to allow operation of the low band mobile service spectrum, such as the 600 and 700 MHz band for satellite use. It remains an open question how 3GPP and other standard bodies might include this proposed use for 5G. To the extent the licensed 5G mobile industry, service providers and equipment vendors, see this service as a threat, they may object, halting progress. They may also be joined by satellite operators who also view this as a threat. Limited action in the standards process would harm the ability for these NGSO MSS systems to form a meaningful part of the 5G ecosystem.

6.4.3 True Hybrid Systems

Another system likely to become more widely available, especially for rural or less densely populated areas to obtain 5G services is hybrid networks. Today these hybrid systems, unlike the systems previously discussed, do not make more efficient use of the spectrum, but utilize two different networks to capture the performance requirements necessary for 5G. For example, today system integrators and others are providing combinations of fixed wireless and GSO satellite to a user. In this hybrid case, the user is provided with a network connection to the fixed wireless and the GSO network, with the network intelligently routing the incoming or outgoing communication over the more appropriate technology. For instance, high capacity, non-latency traffic is routed over the GSO satellite, while latency sensitive traffic is routed over the fixed wireless network. Similarly, as the NGSO mega constellations are deployed, we may see hybrid GSO and NGSO FSS satellite networks with the latency sensitive traffic going over the NGSO satellite and high capacity, non-latency sensitive traffic over the GSO FSS network.

While spectrum efficiency is not a factor here, cost and performance are. By deploying a hybrid network, users in rural areas can obtain higher-quality lower-priced services then by utilizing a single solution. These systems are in early development so whether consumer-priced devices and solutions will be developed are open questions. However, in certain geographically difficult areas or areas with very low population densities, these solutions may make sense to bring advanced 5G services to these users.

It is also important to consider if these hybrid solutions should be encouraged from a spectrum management perspective. In this situation you are using two primary radio services to bring the level of service that perhaps one network using an alternative technology could meet. However, in the case where these hybrid solutions can bring cost-effective services to rural areas where spectrum is more available, this use makes sense utilizing available spectrum to bring users a more robust service. Accordingly, while not a spectrum management issue per se, policy makers may want to look to hybrid services to address issues such as bridging the digital divide in cost-effective manner.

6.5 The Need for a Three-Dimensional Spectrum Model

It is exciting to see the many possible approaches to using spectrum and other tools to make 5G a reality despite the limited available spectrum. However, one concept to consider as we look at the spectrum regimes taking is the need for a three-dimensional spectrum model. A three-dimensional model will become increasingly important as 5G deployments advance on the ground and in the sky. Even if not fully utilized during the 5G deployment timeline, this model will be necessary when the 6G environment starts to unfold, when we are not just connecting everything everywhere, but also advancing sensing technologies.

So, what do we mean when we talk about a three-dimensional spectrum model? This model recognizes that we must improve our evaluation of the potential for harmful interference, and consider all sources of interference. This includes ground interference at varying heights, air space interference within Earth orbits, and beyond into deep space. We would be looking for white or unused space where another operator could utilize the spectrum on an interference-free basis, as well as collective or aggregate interference. However, this model does rely on existing principles seen in spectrum sharing including the increased ability to share in white spaces of the spectrum resource, such as with CBRS.

We are in a very different spectrum world with 5G, and even more likely with 6G. First, we must think about how terrestrial wireless services are providednow and in the future as 5G progresses. Traditionally, wireless base stations have been on high towers transmitting to small and medium devices, whether handsets or points of presence or other towers. However, as more devices are being connected, from watches to cows to refrigerators, new spectrum management models will be needed. In addition, with the use of small cells increasing, you are no longer transmitting from high towers. Now your transmitters can be anywhere,from the median of a highway to a door sill in a classroom. With so many devices transmitting you are increasing your chances for aggregate interference into other radio services [53].

Next, we must look at the airspace, the development of drones, and high-altitude platforms. These systems require communications for TT&C but also to communicate among systems and interconnect with the internet. For example, companies, such as Airbus and Sky-Five, are developing passenger transport systems in the sky for passengers and goods. In our connected world, in-vehicle communications systems are going to be required so that passengers can communicate during their trips.

Further, we must look at near and deep space. First, in near space, we are seeing a significant increase in low-Earth orbit and above satellite systems and space vehicles. For example, Elon Musk's SpaceX System will have upwards of 28,000 satellites. These mega-systems are deployed on the ground, to support massive M2M or IoT communications, or in the sky. With tens of thousands of satellites deployed in a single network talking to and from Earth and with other NGSO and GSO satellites through vast networks of intersatellite links, the interference environment on the ground and in space is going to change. The possibility for aggregate interference in space and on the ground will increase. Compounding this is the real likelihood of increased use of the spectrum for communications for space tourism and sensing in the low-Earth orbit, and for mining, living, and exploration beyond.

Finally, we must consider the impact of deep space including planetary communications. As more countries race for the moon, Mars, and beyond, that means increased spectrum use, with increased potential for harmful interference. For instance, even today, Nokia has announced plans to construct its own 5G network on the moon [54].[3] Combine this with very real scientific missions and the potential for even greater chaos unfolds.

We combine this analysis to determine if there are there periods of time when frequency bands are free from a particular use such that it could be utilized by another, as has been done in CBRS. Could spectrum use be increased if frequencies used in a tower could be used by drones without interference to one another, for a discreet period of time or geographic area?

As regulators look to increase the efficiency of spectrum resource use, including increased spectrum sharing, it will be critical for the analysis to reflect new sharing models that are adopted, including utilizing a three-dimensional spectrum model. This will address very real concerns of the potential for interference for radio services in adjacent or near adjacent bands and territories, and also from different altitudes. To implement this model, regulators will need to seriously consider the total interference environment

3. This is especially important to undertake before too many devices are deployed. History has shown, that it is very hard to pull back widely-deployed devices to address interference concerns once they are in the marketplace.

into and among radio services, on the ground, in the air, and in space. This analysis must also consider aggregate interference which will increasein a 5G world with everything connected to everything else.

To date, some regulators, like the FCC, have been hostile to looking into issues associated with a three-dimensional spectrum model, such as the aggregate interference that can be caused by the widespread deployment of 5G terrestrial devices into satellite network receivers. However, as 5G increases the number of user connections and other radio-based transmitting devices (terrestrial and nonterrestrial), such analysis will be necessary to avoid the potential for harmful interference. Failure to limit harmful interference will actually increase spectrum efficiency. As we have seen in the past, once widespread deployment of radio-based services occurs, even if they cause harmful interference, it is very hard to recall them to fix the problem. This leads to a less efficient use of the spectrum resource.

6.6 Conclusion

As 5G and beyond services are deployed, even if they don't meet current projections, there is no doubt that demand for access to spectrum will increase. While there are many spectrum management models to increase spectrum efficiency, increased use of the spectrum resource calls for new ways of thinking, and evaluating spectrum availability. Increased flexibility must continue to be provided to operators, though there will likely be an increased potential for harmful interference. For example, by allowing a flexible use regime for spectrum acquired on the secondary market, spectrum managers must ensure that such use does not negatively impact the ability of other radio services in the same of adjacent bands from being free from harmful interference. Similarly, allowing new services, such as NGSO systems operating in terrestrial mobile spectrum, opens the door to a wide range of interference concerns never contemplated when the original terrestrial allocations were made.

With that said, there are two imperatives to meet the increasing spectrum demands of 5G and beyond. First, our spectrum access regimes must be more agile and flexible, or innovation will be stalled. If that occurs the benefits of these advanced services will be delayed. Second, to support such flexibility, it is imperative that regulators start to look at new ways to analyze the potential for harmful interference that flexible use combined with 5G and beyond services bring. Accordingly, using new models, such as the three-dimensional spectrum model to examine the interference environment in all dimensions and for all risks, including aggregate interference is critical.

References

[1] Butler, C., "More Ham Radio Spectrum in the 3-12 MHz Needed," *ICQ Amateur,* May 9, 2021, https://www.icqpodcast.com/news/2021/5/9/more-ham-radio-spectrum-in-3-12-mhz-needed.

[2] Ludlum, N., "More Spectrum is Key to America's Future," *CTIA,* April 18, 2019, https://www.ctia.org/news/more-spectrum-is-key-to-americas-5g-future.

[3] Cave, M., "Anti-competitive behaviour in spectrum markets: Analysis and response," *Telecommunications Policy 34 (2010),* 251–261 http://www.elsevierbusinessandmanagement.com/locate/telpol.

[4] Sievert, M., "AT&T and Verizon Are Feeling the Heat of Real Competition," *T-Mobile Newsroom,* https://www.t-mobile.com/news/un-carrier/real_competition_blog.

[5] Cave, M., and Nicholls, R., "The Use of Spectrum Auctions to Attain Multiple Objectives: Policy Implications," *Science Direct*, Volume 41, Issues 5-6, pp. 367–378. June 2017, https://www.sciencedirect.com/science/article/abs/pii/S0308596116302828.

[6] https://www.ida.org/-/media/feature/publications/a/ar/a-review-of-approaches-to-sharing-or-relinquishing-agency-assigned-spectrum/p5102final.ashx

[7] Smith, E., "A Short History of Wireless Spectrum, the Most Complicated Puzzle You've Ever Seen," *Vice,* https://www.vice.com/en/article/mbjpjb/a-short-history-of-wireless-spectrum-the-most-complicated-puzzle-youve-ever-seen.

[8] Tomás, J. P., "Brazilian regulator expects to publish terms of 5G auction in July or August," *RCR Wireless News,* https://www.rcrwireless.com/20210407/carriers/brazilian-regulator-expects-to-publish-terms-of-5g-auction-in-july-or-august.

[9] Tomás, J.P., "France completes 5G spectrum auction," *RCR Wireless*, https://www.rcrwireless.com/20201002/5g/france-completes-5g-spectrum-auction.

[10] FCC, "Secondary Markets Initiative and Spectrum Leasing," ttps://www.fcc.gov/secondary-markets-initiative-and-spectrum-leasing.

[11] Malisuwan, S., et al., "A Study of Spectrum Auction in Various Countries," *International Journal of Science Technology and Management*, Vol. No.5, Issue No. 06, June 2016, http://www.ijstm.com/images/short_pdf/1465919202_P153–166.pdf.

[12] Williams, S., "FCC Poses New Questions to Improve Secondary Market Spectrum Transactions," Fletcher, Health & Hildreth, https://www.commlawblog.com/2019/04/articles/fcc/fcc-poses-new-questions-to-improve-secondary-market-spectrum-transactions/.

[13] Brito, J., "The Spectrum Commons in Theory and Practice." Stanford Technology Law Review, 2006, Available at SSRN. https://ssrn.com/abstract=893964.

[14] Albano, S., "WiFi frequency bands: 2.4 GHz and 5 GHz," *minim*, https://www.minim.com/blog/wifi-frequency-bands-2.4-ghz-and-5-ghz.

[15] Pierce, D., "From Wi-Fi to Bluetooth to 5G, All Your Wireless Is About to Change," *The Wall Street Journal*, https://www.wsj.com/articles/from-wi-fi-to-bluetooth-to-5g-all-your-wireless-is-about-to-change-11548597600.

[16] CEPT, CEPT Report 75 (November 20, 2020), available https://docdb.cept.org/download/135.

[17] Alleven, M., "GSMA calls for governments to license 6 GHz for 5G," *Fierce Wireless*, https://www.fiercewireless.com/operators/gsma-calls-for-governments-to-license-6-ghz-for-5g.

[18] Allevan, M., "6 GHz band pits CTIA against Apple, Google and other tech behemoths," *Fierce Wireless*, https://www.fiercewireless.com/wireless/6-ghz-band-pits-ctia-against-apple-google-and-other-tech-behemoths.

[19] Tarnutzer, B., "A balanced approach to 6 GHz is needed," *GSMA*, August 3, 2020, https://www.gsma.com/spectrum/balanced-approach-to-6-ghz/.

[20] MD7, "The Infrastructure Challenges of 5G—Frequency," February 7, 2020, https://www.md7.com/perspectives/infrastructure-challenges-of-5g-frequency/.

[21] IGI Global, "What is Dynamic Access Spectrum," https://www.igi-global.com/dictionary/fundamentals-of-software-defined-radio-and-cooperative-spectrum-sensing/46200.

[22] "What is Dynamic Spectrum Access?" *Shared Spectrum Company*, https://www.sharedspectrum.com/technologies/dynamic-spectrum-access/.

[23] Lewis, S., "Dynamic spectrum access (dynamic spectrum management)," *Tech Target*, https://searchnetworking.techtarget.com/definition/dynamic-spectrum-access-dynamic-spectrum-management.

[24] Gude, K., "Satellites and Dynamic Spectrum Access," Satellite Evolution Group, 2017, https://www.satelliteevolutiongroup.com/articles/DynamicSpectrumAccess-2017.pdf.

[25] Manner, J., *Global Telecommunications Market Access*. (Norwood, MA: Artech House, 2002).

[26] Freitas, T., et al., "Brazil Orders Cos. to Build Separate 5G for Government Use," *Bloomberg*, https://www.bloomberg.com/news/articles/2021-01-30/brazil-orders-carriers-to-build-separate-5g-for-government-use.

[27] https://www.observacom.org/use-of-the-broadband-spectrum-in-brazil/

[28] Sharwood, S., "After 11 years, Australia declares its national broadband network is 'built and fully operational'," *The Register*, December 23, 2020, https://www.theregister.com/2020/12/23/australian_national_broadband_network_complete/.

[29] Lane, I., "NBN rollout: A success, flop or just middle of the road?" *The New Daily*, https://thenewdaily.com.au/life/tech/2020/06/30/nbn-australia-rollout-2020/.

[30] Broadband Boondoggles, "More About Failed Taxpayer-Funded Broadband Networks," http://munibroadbandfailures.org/.

[31] Gangitano, A., "White House getting pushback on possible government-owned 5G network," *The Hill*, October 26, 2020, https://thehill.com/policy/technology/522698-white-house-getting-pushback-on-possible-government-owned-5g-network.

[32] Mehlman, B., and Boucher, R., "Why a government-owned 5G network is still a bad idea," *Roll Call*, October 21, 2020, https://www.rollcall.com/2020/10/21/why-a-government-owned-5g-network-is-still-a-bad-idea/.

[33] Dano, M., "Industry debates 'the art of the possible," in 5G spectrum sharing," *Light Reading*, February 10, 2021, https://www.lightreading.com/ai-automation/industry-debates-the-art-of-possible-in-5g-spectrum-sharing/d/d-id/767302.

[34] Manner, J., *Global Telecommunications Market Access*. (Norwood, MA: Artech House, 2002).

[35] van Burken, R., "A Federal Government-Owned 5G Network Would Be A Disaster," *Reason Foundation*, November 20, 2020, https://reason.org/commentary/a-federal-government-owned-5g-network-would-be-a-disaster/

[36] Danel, E., "Wi-Fi 6E Spectrum in 6 GHz Band," *LitePoint*, https://www.litepoint.com/blog/an-introduction-to-wi-fi-6e-spectrum/.

[37] "Impacts of the LightSquared Network on Federal Science Activities: Hearing before the Committee on Science, Space and Technology, 112th Cong.," 2011. Available at https://www.govinfo.gov/content/pkg/CHRG-112hhrg68227/pdf/CHRG-112hhrg68227.pdf.

[38] Weimer, B., and Brooks, D., "FCC Proposes to Grant DISH's Wish, *JD Supra*, https://www.jdsupra.com/post/contentViewerEmbed.aspx?fid=4a4a735f-ec0e-433e-8497-fa6258dfcff1.

[39] Mullin, S., "FCC Proposes to Grant DISH's Wish," March 23, 2012, https://www.fcclawblog.com/2012/03/articles/fcc/fcc-proposes-to-grant-dishs-wish/.

[40] FCC, "In the Matter of Service Rules for Advanced Wireless Services in the 2000-2020 MHz and 2180-2200 MHz Bands, Report and Order," December 17, 2012 https://docs.fcc.gov/public/attachments/FCC-12-151A1.pdf

[41] https://en.wikipedia.org/wiki/Ligado_Networks

[42] Reardon, M., "The Pentagon's fight to kill Ligado's 5G network," CNET, May 13, 2020, https://www.cnet.com/tech/mobile/the-pentagons-fight-to-kill-ligados-5g-network/.

[43] de Selding, P., "Inmarsat, citing SpaceX delays, to miss European deadline for aeronautical broadband service," *Space News*, https://spacenews.com/inmarsat-citing-spacex-delays-to-miss-european-deadline-for-aeronautical-broadband-service/. Morris, A., "Echostar buys struggling MSS Operator Solaris Mobile," *Fierce Wireless*, https://www.fiercewireless.com/europe/echostar-buys-struggling-mss-operator-solaris-mobile.

[44] Hill, J., "EU Court Rejects Eutelsat, Viasat Challenge to Inmarsat EAN Authorization in France," *Via Satellite*, April 15, 2021, https://www.satellitetoday.com/business/2021/04/15/eu-court-rejects-eutelsat-viasat-challenge-to-inmarsat-ean-authorization-in-france/.

[45] Rainbow, J., "Ligado Networks passes 3GPP review for 5G plan," *Space News*, June 22, 2021, https://spacenews.com/ligado-networks-passes-3gpp-review-for-5g-plan/.

[46] https://www.commsalliance.com.au/__data/assets/pdf_file/0015/60315/SSWG-Ensuring-the-Future-of-Australian-Satellite-Services.pdf.

[47] https://www.rsm.govt.nz/assets/Uploads/documents/consultations/2020-1710-2300-mhz/omnispace-submission-re-planning-frequency-bands-1710-2300-mhz.pdf.

[48] FCC, "In the Matter of AST&Science LLC Petition for Declaratory Ruling Granting Access to the U.S. Market for a Non-U.S.-Licensed Non-Geostationary Orbit Satellite Constellation," November 19, 2020.

[49] https://apps.fcc.gov/els/GetAtt.html?id=222501&x=.

[50] FCC. "Updates Concerning Non-Geostationary, Fixed-Satellite Service Systems and Related Matters," Federal Register 86, February 26, 2021. 11642. https://www.federalregister.gov/d/2021-04028.

[51] Milne, A., "Common Sources of Interference," *RF Venue*, https://www.rfvenue.com/blog/2014/12/14/common-sources-of-interference.

[52] https://spectrum.ieee.org/telecom/wireless/electronic-noise-is-drowning-out-the-internet-of-things, Nokia, "Nokia selected by NASA to build first ever cellular network on the Moon," https://www.nokia.com/about-us/news/releases/2020/10/19/nokia-selected-by-nasa-to-build-first-ever-cellular-network-on-the-moon/.

[53] FCC, "Use of Spectrum Bands Above 24 GHz For Mobile Radio Services, et al.," *Fifth Report and Order*, 34 FCC Rcd 2556 (3), (2019); FCC, "Use of Spectrum Bands Above 24 GHz For Mobile Radio Services, et al.," *Report and Order and Further Notice of Proposed Rulemaking*, 31 FCC Rcd 8014, 8165–67 (2016).

[54] FCC, "In the Matter of Review of Part 15 and other Parts of the Commission's Rules," *Notice of Proposed Rule Making and Order*, 17 FCC Rcd 17003 (22) (2002).

7

Enabling the Use of Additional Spectrum for 5G

7.1 Introduction

This chapter explores several ways many governments and spectrum managers are using innovative approaches to make additional spectrum available to support 5G and beyond services. The goal of these regulatory approaches is increasing the efficient use of the spectrum, getting the resource into the hands of operators, and ultimately to end-users. These innovative approaches include novel auction designs along with more traditional license conditions. With government focus on solving the digital divide, some countries, like France, are imposing build-out conditions on operators. These build out conditions require operators to utilize the spectrum they buy at auction for 5G service in rural areas, actually build out and provide service.[1] Due to the differences associated with use of the spectrum for 5G used by satellite networks, we cover critical areas like blanket licensing the sharing of spectrum between NGSO networks.

These approaches are not static, are continuously improved, and new approaches are constantly being considered. We end the chapter with a discussion of the challenges countries are facing developing and implementing additional novel approaches for more spectrum access to meet growing user demands by all technologies for 5G and beyond.

7.2 Getting Spectrum Into the Hands of Operators

7.2.1 Secondary Markets and Use It or Lose It

Since the 2000s, countries have been enabling secondary use of spectrum to enable more efficient use of the spectrum resource, and to get additional services out to users [1]. No matter how efficiently spectrum is allocated and assigned, changing conditions of supply and demand mean that operators may have temporary and long-term spectrum that remains unused. Secondary markets allow spectrum to be transferred, or leased to other more efficient users while leaving the primary licensee responsible for meeting license obligations, including license build-out requirements. While secondary markets are available in some countries, here we focus on two approaches: 1) the United States FCC's secondary market rules, and 2) the use it or lose it policies of some countries.

7.2.2 The FCC's Secondary Markets Approach

The FCC's secondary market rules, around since the early 2000s, have the goal of encouraging secondary market transactions through efficiently distributing spectrum [1]. Under the FCC's secondary market rules, licensed terrestrial wireless spectrum can be transferred or leased to another entity pursuant to the FCC's

1. A constant concern for regulators is that while terrestrial mobile operators have licenses that cover broad geographic and rural areas, often it takes significant time for their networks to be built out in these more rural areas. Regulators may be hesitant to force build-out where service may not be economically viable.

secondary market rules.[2] Spectrum leases allow a lessee to use a licensee's spectrum without assigning the license to the new user. The arrangement is made between the parties and may involve the leasing of any amount of licensed spectrum, in any geographic area or site encompassed by the license, for any period of time during the term of the license authorization. The transfer process is accomplished through FCC rules on how to lease the spectrum with the lessor remaining responsible for meeting FCC license obligations.[3]

While the FCC was hopeful that its secondary markets approach would result in a fair number of operators participating, that has not been the case [2]. There are many possible reasons for this including lack of demand, competitive concerns, and the regulatory burdens associated with the leases. In addition, the limited interest may reflect the lack of commercial viability of the available markets. Another possible reason for the lack of use may be that the FCC's rules currently are limited to geographic leases; not spectrum uses.

To increase secondary market leases with 5G on the horizon the FCC has a pending "rulemaking" procedure to allow partition and disaggregation of spectrum on the secondary market. Partition is the geographic division of a spectrum license. All licensed spectrum in the partitioned area would be assigned to the assignee who could then transfer various partitions. Disaggregation involves assigning certain blocks of spectrum (e.g., channels or bands of spectrum) in a licensee's entire geographic area associated with a license. The FCC's proposed rules allow licensees to partition and disaggregate a license, thus providing licensees with considerable flexibility on how they want to sell their geographically licensed areas and spectrum [3]. It will be interesting to see if this increases the use of the FCC's secondary market rules as other countries, like the United Kingdom, have adopted similar such changes, but not seen much improvement.

---

2. Under the FCC's rules, spectrum cannot be leased for satellite use.
3. Other countries have also authorized secondary markets. For instance, in Slovenia's recent 5G spectrum auctions, the government allows secondary use of the spectrum with the regulator's approval. Like in the United States regime, as discussed below, there have been few secondary market agreements being entered into.

7.2.3 Use It or Lose It

While not a new idea, conditioning 5G licenses on a *use it or lose it* basis continues to be one way spectrum managers can increase spectrum use and efficiency. The use it or lose it condition incentivizes operators to make use of the spectrum or face the risk of losing access to the resource. This approach has been implemented in Europe through their mobile spectrum auctions for some time. Many European auctions came with use it or lose it clauses that obligated operators to build a minimum number of base stations or face sanctions from fines to loss of spectrum [4]. These countries use this approach to encourage spectrum use and allow a dynamic secondary market for spectrum. The United States has implemented its own use it or use it rule for certain wireless services licenses.

These use it or lose it rules are still in use and apply to some 5G spectrum licenses. For example, Brazil recently announced a plan to implement an approach for spectrum (it is uncertain if this is only for terrestrial spectrum or also for nonterrestrial and/or for fixed or mobile services), where if licensed spectrum is not available in five years, it would be made available for a secondary use [6].

However, we are now seeing use it or lose it conditions being used to encourage build out in more rural areas to help bridge the digital divide. France is a good example of this approach with its 2020 auction for 5G spectrum where it imposed a mix of requirements to ensure spectrum efficiency and service to rural areas in its 3.4-3.8 GHz auction [7]. Such conditions included:

- Launch of service in at least 2 cities by end of the auction year;
- Having each carrier deploy 3,000 sites by 2022, 8,000 sites by 2024, and 10,500 sites by 2025;
- 25% of 3.4-3.8 GHz band sites in the last two stages must be located in sparsely populated areas, targeting economic activity, notably manufacturing, excluding major metropolitan areas;
- By 2022, at least 75% of cell sites must be capable of providing speeds of at least 240 Mbps at each site;

- The obligation for the carrier to deploy 5G to provide coverage in roadways across France.

While lose it or use it conditions can achieve their goals, regulators are often lax at actual enforcement. There are several instances where operators have missed their build-out obligations, but instead of losing the spectrum, have received waivers and/or extensions [8]. A good example of this was the DISH Networks extension for its spectrum in the AWS-4 band [9]. However, in this case, obtaining waivers/extensions have come out of the cost of additional commitments, including new build-out requirements imposed on the operators. A key reason why such extensions/waivers have been granted is the concern by regulators that if the spectrum is taken away from one operator and given to another, the new licensee may face the same economic challenges on build-out as the original licensee.

As previously discussed, use it or lose it is being used at the ITU for NGSO FSS satellite networks, with the possibility that failure to deploy a committed percentage of satellite, may lead to loss of spectrum. It is too early to tell if this approach will result in greater use of the spectrum by the network operator. However, it will free up spectrum earlier during the build out for other users, which should result in greater spectrum use over the long term.

While the application of use it or lose it provisions asincentives to utilize spectrum, it remains unclear if they are an effective tool at incentivizing greater use of the spectrum resource.

7.2.4 CBRS and Other Approaches

Another recent novel approach to make spectrum more widely available is the CBRS approach undertaken by the FCC in the 3350-3.7 GHz band (the CBRS frequency band). This band was already in use by two incumbent users that left the band underutilized; 1) FSS (space-to-Earth) Earth stations; and 2) DOD (military) radar systems [10]. Because of the limited use by the incumbent users and the value of the CBRS frequency band for 5G, it was targeted by technologists, the U.S. government, and the wireless industry as an important mid-band spectrum for 5G deployment. The challenge that U.S. spectrum managers faced was how to enable 5G use of this band while enabling continued use by the incumbents. The FCC was quite successful in resolving the challenge, and the CRBS auction ultimately generated U.S. $4.5 billion [11].

CBRS utilizes a three-tier approach to spectrum access. The space is shared between the incumbent access user, a priority access licensee (PAL), and a general authorized access (GAA) user [12]. As the incumbent access users, the U.S. military radars and satellite services in the CBRS band are fully protected at all times. PAL tier users operate in licensed auctioned spectrum protected against GAA use but must protect incumbent access users. GAA users are unlicensed users that can operate in nonauctioned spectrum with the obligation to protect both the PAL licensees and the incumbent users.

To enable sharing among the different tiers, the FCC created an automated frequency coordinator known as a spectrum access system (SAS) managed by a private entity contracted to by the federal government [13]. To use CBRS spectrum, each user must individually request and be assigned a frequency band by the SAS programmatically. The SAS performs the calculation for RF density and channel availability using terrain and radio propagation data before authorizing the request. When this spectrum is no longer required, it is freed up for use by other requestors. By using the SAS, the PAL licensees are guaranteed to have interference free communications.

We are still at a very early stage with CBRS and the use of the SAS. However, if successful, this effort will provide a valuable model that canbe used by other countries to enable sharing in a way that protects existing uses while allowing a more robust use of a frequency band. One remaining issue is whether this approach is appropriate where incumbent use is growing because of the auctioned nature of some of the spectrum. In this case, auction licensees would likely raise concerns if they were somehow limited in access to spectrum they may need. But where the incumbent use is static or declining, the CRBS model is a very good one for spectrum managers to consider.

7.2.5 Incentive-Based Spectrum Management

7.2.5.1 The U.S. 600 MHz Incentive Auction

The 600 MHz band has been viewed as prime low band spectrum for 5G. It has very good propagation characteristics meaning its signals can travel long distances and permeate through some structure. However, long-term broadcast incumbents made it impossible to clear in the United States. With U.S. 5G operators clambering for more spectrum, the FCC developed the world's first incentive auction for use in this band. This auction was viewed as a resounding success for the FCC.

This auction comprised three separate but interdependent parts [14]. Two auctions and a middle *repacking process*. The first auction was the *reverse auction* which determined the price at which broadcasters would voluntarily relinquish their spectrum usage rights. The second was a *forward auction* that determined the price companies would pay for flexible use wireless licenses.

The repacking process involves reorganizing and assigning channels to the remaining broadcast television stations to create contiguous blocks of cleared spectrum suitable for flexible use [14]. The auction design recognizes there will still be broadcasters who do not participate in the auction. The vast majority of these stations were assigned channels in the TV band, while in markets where the post-auction TV band is not large enough to accommodate every station, stations may be assigned a channel in the wireless band.

The reverse and forward auctions were implemented in stages with each stage consisting of a reverse auction and a forward auction. Prior to the first stage, the initial spectrum clearing target was determined. Through the pre-auction application process broadcasters indicated their willingness to relinquish spectrum usage rights at the opening prices.

Based on broadcasters' collective willingness, the initial spectrum clearing target was set at the highest level possible (up to 126 megahertz of spectrum) without exceeding a predetermined national aggregate cap on the interference between wireless providers and TV stations (impairments) created when TV stations are assigned to the wireless band. With this approach the auction system established a band of wireless spectrum that was generally uniform in size across all markets. The goal of the reverse auction bidding process was to determine the total incentive payments to broadcasters required to clear that amount of spectrum.

The forward auction bidding process followed the reverse auction bidding process. When the final stage rule was satisfied, the forward auction bidding continued until there was no excess demand, and then the incentive auction closed. If the final stage rule was not satisfied, additional stages were run, with progressively lower spectrum targets in the reverse auction and less spectrum available in the forward auction. The final stage rule was a set of conditions that must be met to close the auction at the current clearing target. Failure to satisfy the rule would result in running a new phase at the next lowest clearing target.

The final stage rule has a reserve price with two components, both of which had to be satisfied. The first component required that the average price for low impairment licenses in the forward auction meet or exceed $1.25 per MHz-pop at a 70 MHz cleared benchmark. Alternatively, if the spectrum clearing target at a particular stage is greater than 70 MHz, then the first component would be met if the total proceeds of the forward auction exceed the product of $1.25 per MHZ/pop x 70 MHz x the total number of pops for the high-demand partial economic areas (PEAs) with at least one Category 1 block in this stage. This alternative formulation allowed the auction to close if the incentive auction repurposes a relatively large amount of spectrum for wireless uses, even if the price per-MHz-pop is less than the benchmark price.

The second component of the final stage rule required that the proceeds of the forward auction be sufficient to meet mandatory expenses set forth in the Spectrum Act. If the requirements of both components of the reserve price are met, then the final stage rules were satisfied [15].

On March 30, 2017, the auction satisfied both conditions of the final stage rule. The auction raised $19.8 billion USD making valuable low band spectrum available to multiple mobile wireless licenses for 5G [16].

The United States has completed the repacking effort and today 600 MHz continues to be deployed and used for 5G. This model may be attractive to other countries as a way to clear broadcasters from the 600 MHz band to put this spectrum to higher use, while preserving the ability of broadcasters to operate on a more efficient basis.

7.2.5.2 C Band (3.4 GHz to 3.8 GHz)

The C band is a very attractive band to terrestrial mobile 5G operators because of its propagation characteristics and volume of bandwidth. While satellite operators have been using this band for years, it is unclear if the use justified access to the full complement of C band spectrum. To this end, a number of countries have been focused on making this band available for 5G, especially since WRC-15 allocated this band for 5G mobile services, while providing limited protections to FSS operators. However, the resistance of the satellite industry and the important users on this band (e.g., broadcasters) has made it very difficult to relocate the FSS users to another band to free up adequate spectrum for 5G. Nonetheless, with increased pressure to make more spectrum available for 5G mobile terrestrial in the United States it became inevitable that the United States would find a way to obtain access for 5G in this valuable band.

This pressure led Intelsat, one of the largest global satellite operators, to propose to the FCC with Intel, a structure whereby the satellite operators would be compensated for their costs of transitioning from the band, moving them to a smaller amount of shared spectrum [17]. This compensation would include relocation costs including the construction of new satellites.

This approach was first widely fought by the satellite industry and others, but ultimately it formed the basis for the FCC's C band rules including relocation of the C band satellite operators and auction of this spectrum for 5G. Specifically, the FCC was able to free up 300 MHz of the C-Band for 5G, 280 of that to auctioned, and 20 MHz to be used as a guard band between wireless users and the incumbent satellite operators. They will use the remaining 200 MHz to continue delivering network programming to broadcast and cable clients (and others to be compensated for their transition). In addition to covering costs of relocation, the FCC approved up to $9.7 billion in incentive payments to those satellite carriers for exiting the spectrum early, which was widely accepted. The auction resulted in $78 billion in auction revenues [18].

Despite a successful transition in the United States, this approach is not as viable in other parts of the world. Especially in developing countries the satellite industry operating in the C band is put to greater use than in the Unites States. Therefore, there is greater opposition to transition away from the C-band. Such opposition may also reflect that few countries will be able to develop a regime that provides the compensation to incentivize the satellite operators to transition to a smaller portion of the frequency bands.

7.3 Access to the 28 GHz Band for 5G Terrestrial Services

The 27.5-30 GHz band (28 GHz band) has been utilized for the fixed satellite service for many years for gateway, and for user terminals. In parts of the band, such as 27.5-28.35 GHz, there has been an allocation for the fixed and mobile services on a coprimary basis in countries around the world. Since the FSS and fixed service were able to successfully share this band, many countries, including the United States, instituted this sharing scenario (i.e., between the fixed service and FSS operations).[4]

Due to light usage of this band the terrestrial mobile industry, led by mobile operators and equipment vendors, targeted the entire 28 GHz band leading to WRC-15 to add it to the WRC-19 Agenda as a candidate band for FSS. One assumes that the terrestrial mobile proponents and their allies knew it was a long shot to have the entire band identified for IMT, and had planned to fall back to just 27.5-28.35 GHz, a band where FSS gateways (not user terminals) operated primarily. This meant that the sharing environment was easier to address.

To shore up support for this identification, the terrestrial mobile industry worked hard to get countries including the United States on-board. The U.S. started a proceeding to make the 27.5-28.35

4. The rest of the 28 GHz band was heavily used since the mid-2000s for satellite broadband use which is seen as key to provide broadband services in rural and remote parts of the globe, including Africa and Latin America.

GHz band available on a primary basis for 5G terrestrial, allowing satellite use only in the most rural and remote parts of the country.

While the mobile industry lobbied for their position, the satellite industry joined together to fight this effort internationally. They were successful due to overwhelming support from the developing world, and the global recognition of the role satellite broadband plays in solving the digital divide. At the end of WRC 2015 there were only 4 countries who vocally supported studying the 27.5-28.35 GHz band for IMT. Instead, the focus turned to the study of portions of the band from 24 GHz to 27.5 GHz, and the bands above 30 GHz to be identified at WRC-19 for IMT [19].

Despite this loss and the lack of the ability to ensure global harmonization of the 28 GHz band for IMT, the advocates for 28 GHz did not relent. Instead, the United States and other countries including Canada, and Chile moved forward making the 27.5-28.35 GHz band for terrestrial 5G at the expense of satellite [20, 21]. For example, in the United States, rules were adopted that provided terrestrial 5G with primary use of the band, and limited use of the band for FSS to a small number of gateway Earth stations located in the most rural and remote portions of the United States [22, 23].

The terrestrial mobile industry is actively continuing this effort on a country by country basis despite the possibility for global harmonized use of this band. The result of these efforts by the terrestrial mobile industry, and the governments who have aligned with them on 28 GHz, is to limit the spectrum available for FSS services in these countries. At the same time spectrum terrestrial use will increase which can limit the availability of broadband and 5G satellite service in rural and underserved parts of the world.

7.4 Accessing Higher Frequency Bands (Above 30 GHz)

The bands above 30 GHz (i.e., 37.5-40 GHz, 40-42 GHz, 47.2-50.2 GHz, and 51.4-52.4 GHz) have been lightly used by commercial interests despite long-standing spectrum allocations. For example, FSS has been allocated in the 37.5-42 GHz bands and the 47.2-50.2 GHz and 50.4-51.4 GHz bands for years, but has only recently been seriously considered for use by FSS.[5] This is partly due to the fact that the technology did not allow such access. However, recently, satellite technology, like the terrestrial 5G technology, has evolved to allow use by the terrestrial mobile and fixed satellite service operators (despite limited equipment for either in these bands today). In the past few years there has been a struggle between the satellite and terrestrial operators for access to this spectrum at both the ITU and in domestic markets.

The actions at WRC-19 laid the groundwork for use of the above 30 GHz bands for 5G terrestrial by identifying the spectrum for IMT and the countries moving forward with implementation of this decision. However, not all countries are making these bands fully available for terrestrial operators. For example, the United States has addressed sharing with the FSS in these bands in a somewhat balanced manner setting aside 2 GHz by 2 GHz of the 40-42 GHz band and the 48.2-50.2 GHz band for FSS use on a primary basis, and 5G mobile use on a secondary basis. The U.S. then made a primary allocation in the 37.5-40 GHz and 47.2-48.2 GHz bands for 5G terrestrial, with FSS being secondary. Other countries are examining similar plans that would reserve some of these bands for satellite use. Due to the propagation characteristics of the above 30 GHz bands, it makes sense to carefully examine how much of these frequency bands should be made primary for terrestrial 5G versus satellite use.

7.5 Set-Asides for Industrial/Private Networks

Throughout the long history of spectrum management, a number of industries or aligned users have argued for spectrum set-asides. With spectrum increasing in value, this has only grown with time. These groups include the public safety sector, utilities, railroads, and more recently to specific applications, such as IoT. These industries and users are pushing harder than ever for access to dedicated spectrum in the 5G world.

5. As of the date of publication of this book, only experimental use of these bands by the satellite industry, although several satellite networks are planned for these bands, including Hughes' Jupiter 3, which will be launched in the second half of 2022.

Internationally, the push for set-asides is continuing to increase. For instance, on the WRC-23 Agenda, Agenda Item 1.18 was advanced by two small satellite operators and their administration for allocating spectrum for commercial IoT MSS purposes. Similar efforts to advance dedicated spectrum for IoT uses has been advanced at the ITU as well. There are concerns about this approach since it would limit the use of a radio allocation to a specific use or application, which may or not make sense. This is contrary to an allocation of spectrum to a radio service, such as mobile satellite service, where there is freedom for a variety of uses that meet the technical parameters contained in the Radio Regulations. Therefore, while it has been argued that at the ITU allocating spectrum to a specific application will give that use an opportunity to grow in an otherwise crowded spectrum environment, in fact this may result in leaving valuable spectrum fallow if the advocates fail to use it.

Spectrum set-asides are also utilized domestically and continue to be advanced. For example, countries including the United States have traditionally set aside spectrum for public safety use. Again, this may seem like a good idea, but it does not always lead to the most efficient use of the spectrum resource. In the United States, for example, portions of the 700 MHz band assigned for broadband public safety use, and the 4.9 GHz band with a similar use restriction were widely unused. Only in the past few years, has the 700 MHz band been used more widely for public safety broadband in the United States, largely because the band is now operated by and shared with AT&T [24, 25]. While the FCC continues to look for innovative ways to increase use of the 4.9 GHz band for public safety (through leasing options) this band remains largely unused [26, 27]. While questions remain whether spectrum set-asides make sense, other countries are continuing with spectrum set asides. A recent example is India making 5 MHz of spectrum in the valuable 700 MHz band available for railways [28]. Another example is Germany setting aside 100 MHz of the 3.4-3.8 GHz band for vertical licenses [29].

An emerging trend with 5G is making spectrum available for small companies or for private networks [30]. For example, Finland has actually issued private licenses as part of its recent 5G spectrum auctions, and countries like Malaysia are issuing private network licenses on a first come first served basis for uses such as building 5G networks at ports [31]. It is likely if these initial efforts are successful, other countries will follow.

So, despite evidence that set asides and making limited amounts of spectrum available for private networks are not always spectrally efficient, countries continue using set-asides to meet certain needs [32]. It remains uncertain whether making spectrum available for private or limited networks will be successful, especially with the cost of constructing and operating a network. It seems the better approach might have been for these entities to utilize leased spectrum, but that has been largely unsuccessful for a number of reasons. Spectrum managers will need to stay vigilant to ensure that these networks are built, utilize the spectrum resources assigned, or reassign the spectrum in a manner that best supports 5G and beyond.

7.6 Local Licensing

Another option to increase efficiency and the use of spectrum resource terrestrially is through licensing. Germany, for example, made about a quarter of its 3.5 GHz band available for 5G use by local licensees. Germany has devoted one quarter of its prime 3.5 GHz band 5G spectrum for hyper-local licenses, as well as the 24.25-27.5 GHz band. At the time of this book, it was also considering local licenses in the 3.7-3.8 GHz band. In addition, the United Kingdom's Ofcom has also introduced local licensing in parts of the 3.8-4.2 GHz, the 1800 MHz, the 2.3 GHz, and the 26 GHz bands. This spectrum is available on a first come, first served basis, and only cost-based administrative fees are charged. These regulators view local licensing as another way to get spectrum in the hands of operators for 5G, and ultimately to users, in innovative ways with services more responsive to local needs. It is still too early to see how effective local licensing is to increase spectrum efficiency, especially since the license areas are relatively small.

7.7 Increasing Access to Government Spectrum by Commercial Operators

As commercial spectrum becomes scarce, there is a scramble in developed countries to obtain access to government spectrum. This is best illustrated by recent actions in the United States. Besides the CBRS auctions, the United States terrestrial wireless industry continues to push for access to additional government spectrum. To this end, the FCC recently announced that the 3.45-3.55 GHz band would be made available for commercial mobile 5G use [33]. The band used by certain radar systems critical to DOD operations, and as part of the framework adopted for auction of this band, will be transitioned to different frequency bands.

Conversely, the UK has a different view on government spectrum use. Unlike other countries the UK charges administered incentive pricing to ensure efficient use of the spectrum resource, [34, 35]. In most other countries, fees are only paid to cover actual costs of spectrum management administration.

The demand by commercial operators to obtain access to government frequency bands will likely continue. However, some of these government agencies, such as the U.S. DOD, have important uses in these bands, and the transition of the government out of the bands requires a careful balance by spectrum managers.

7.8 Regional and Subregional Harmonization and Implementation of 5G Bands

Traditionally, the ITU has been the body where frequency bands are harmonized on a regional and global basis for different spectrum use including for 5G services. However, all the ITU does in the identification of spectrum is place a guidepost on spectrum available for use; it does not actually implement use. This is where a regional or subregional harmonization *and implementation* of 5G bands approach makes sense.

This harmonization effort was seen in recent European Union action adopting a harmonized approach to 5G spectrum in three frequency bands: the 700 MHz, the 3.6 GHz, and the 26 GHz bands, requiring these bands be made available for 5G by 2020

[36]. This appeared to encourage quick implementation of 5G networks across the EU, because of the binding nature of EU directives. However, even with these tools available, very few countries have met the deadline [37]. This is largely due to countries not believing this spectrum availability is a high priority compared to other governmental functions or due to low demand. Despite the claimed demands for spectrum for 5G, countries are concerned about the lack of entrants. To incentivize operators Spain is lowering minimum bid prices and extending license terms [38]. This failure to harmonize and implement services underscoreshow sovereignty directly impacts the spectrum management regime.[6]

Similar harmonization efforts are being made in various other regions of the world. In terms of making the 27.5-28.35 GHz band available for 5G in the Americas, the United States—an early adapter made this band available for 5G in 2017—encouraged its North America and Americas regional colleagues to do the same. North America implementation of this band for 5G was particularly important to the United States due to shared borders with Canada and Mexico. Canada followed with harmonization creating its own rules to make this band available for 5G [39]. However, Mexico received significant opposition from the satellite broadband industry, and due to its national interest in preserving spectrum for broadband to more rural areas on the country, has not moved to make this band available for 5G.

Despite the failure to harmonize services with Mexico, the United States, and Canada are actively working together to push CITEL-wide (the Americas region) implementation. To date, though significant outreach has been made on a bilateral and multilateral basis by these countries and the wireless industry, only a handful of countries in the Americas have the 27.5-28.35 GHz band available for 5G, Chile being one. These countries view the incumbent FSS use (satellite broadband) as an important part of their infrastructure to provide users in rural areas with broadband. However, if 5G deployment in these bands is successful in the United States and Canada, providing better service to rural areas than traditional mobile services, it may provide the push needed for some

6. The only successful example of EU-wide spectrum management was in the context of MSS/CGC in the mid-2000s.

other CITEL region countries to adopt this regime. This decision will have to be carefully balanced against the use of the band for 5G and broadband satellite services, currently used to support rural users across Latin America.

For 5G and beyond services for all to be successful, it is critical for regulators to work together to harmonize the use of the spectrum resource as much as possible. However, harmonization depends on what use is made by oincumbents in the frequency bands, and whether countries can agree on the appropriate balance between the incumbent and new users seeking to have the band harmonized.

7.9 The Challenges of Increased Use by Satellite Systems

Due to the regional and global nature of satellite systems, there are additional and somewhat different issues associated with getting spectrum out for use for 5G. Below we touch on two of them; the need for blanket licensing, and how to handle the assignment of spectrum for NGSO satellite systems.

7.9.1 Blanket Licensing of User Earth Stations

With the proliferation of satellite Earth stations, the regulatory and spectrum management regime has evolved. It must continue to award efficiencies as satellite networks become an integral part of our 5G ecosystem. Satellite networks will continue to provide 5G service to users, their homes, and will increasingly provide services on moving platforms, oftentimes using identical user terminals. Once such a terminal is authorized, the potential for harmful interference from additional terminals has diminished dramatically. This has led some countries to implement a blanket licensing regime for very-small-aperture terminals Earth stations (also known as VSATs). Under such a regime, the licensee is authorized for a significant number of identical terminals, and does not have to coordinate terminals at each location. In addition, only a single fee is charged for these terminals, ideally based solely on administrative costs, which in turn reduces the cost to end users.

As satellite becomes an increasing part of the 5G solution, it is critical for governments to adopt blanket licensing for VSATs. This will accommodate more users faster and cheaper, resulting in greater spectrum efficiency. This is important if satellite operators are going to be an important part in deploying 5G, and needs to be extended to a wide variety of satellite devices [40].

7.9.2 Assigning Spectrum to Meet Increased NGSO Demand

As previously discussed, we are in an exciting time in the satellite industry as we see multiple proposed satellite networks including mega constellations. However, because of the international nature of satellite systems, many countries, including the United States [41], do not auction satellite spectrum. In the few cases where it has been done, such as Brazil, it has not been successful over the long term [42].

Traditionally, for FSS GSO systems, first in time, first in right has worked since the system is operating at a single location (only one system can be authorized.) However, NGSO satellite systems are different. In most countries, first in time, first in right continues to govern for their licensed systems. However, the United States has taken a different view of systems it licenses as well as with those who seek market access. The United States believes spectrum can be shared among various systems (as well as with GSO systems), with them operating in different configurations based on coordination between one another. Accordingly, instead of assigning specific bands to each operator, the FCC has created an incentive for successfull coordination. It has imposed a condition on the licensed NGSO operators for common frequency bands to successfully coordinate among themselves or the spectrum will be divided equally among the operators [43].

The U.S. approach has pros and cons. The biggest pro is that since several of the systems may be speculative, it may force operators not planning to deploy their systems to surrender their licenses. However, this is also bad since it encourages speculation.

The United Kingdom has adopted a similar approach which relies on cooperation among the NGSO systems to share spectrum. What is clear with both the US and UK policies is that it does not necessarily result in the most efficient use of the spectrum resource. Instead, it favors those with money and who have the longest staying power. In the U.S. not one of the entities could launch and operate an economically viable system with the spectrum assigned; it is a fiction. What is true of both approaches is that there is significant uncertainty for all the systems involved. While less troubling for the larger, well-financed constellations, such as Elon Musk's Starlink, and Amazon's Kuiper, other satellite systems relying on public and other financing face uncertainty that could negatively impact their staying power.

7.10 Remaining Challenges in Implementing New Spectrum Management/Assignment Regimes to Accommodate 5G and Beyond

Although governments have made significant progress in making spectrum available for 5G, there are still many challenges to be addressed. These include:

1. *Cost:* Many governments simply do not have the financial resources to develop and/or implement new approaches to increasing spectrum efficiency. This includes resources needed for personnel, software, and research costs.
2. *Technology sophistication:* Many governments do not have the skilled workforce available to develop and implement complex auctions and other spectrum efficiency schemes.
3. *Focus on attracting bidders:* For developing countries in particular, it is hard to attract licensees for areas that are rural or might not have the most attractive economics or return on investment. Thus, imposing build-out requirements and other conditions on spectrum licenses could further discourage interest.
4. *Strength of incumbents:* Many of the incumbents in frequency bands most attractive for 5G have politically powerful incumbents, like broadcasters. Even with schemes like incentive auctions, it is difficult in some

countries to adopt such approaches because of political influence.

5. *Anti-competitive actions:* Currently, several countries have limited competition in the terrestrial and/or nonterrestrial wireless markets. Therefore, powerful incumbent operators are often roadblocks to governments that seek to make additional spectrum available.
6. *Trust:* In some countries, where the regulatory process and/or the government is not stable or transparent, potential market entrants may be hesitant to enter a market for spectrum access using an untried approach.

Regulators, spectrum managers, and governments that seek to make spectrum available for 5G and beyond, face many challenges and competing goals. These includehow to protect incumbents while also encouraging deployment of 5G on a timely basis over a wide geographic area. As spectrum managers set forth plans, they must carefully balance these challenges and competing goals to determine what makes sense, including using some of the approaches above to determine which one best achieves the goals for making sufficient spectrum available for 5G services.

References

[1] OECD (2005-04-20), "Secondary Markets for Spectrum: Policy Issues," *OECD Digital Economy Papers*, No. 95, OECD Publishing, Paris, http://dx.doi.org/10.1787/232354100386.

[2] New America, "6/3 FCC Comments on Secondary Markets with Public Knowledge," https://www.newamerica.org/oti/wireless-future-project/legislativeregulatory-filings/63-fcc-comments-secondary-markets-public-knowledge.

[3] 47 C.F.R § 101.1111 (2021)

[4] See, Bode, K., "Do We Need 'Use it or Lose it' Spectrum Rules? - Sanford Bernstein Analyst Makes a Very Good Case," *DSL Reports*, http://www.dslreports.com/shownews/Do-We-Need-Use-it-or-Lose-it-Spectrum-Rules-114921.

[5] 47 C.F.R § 1.953 (2021)

[6] Reuters, "Brazil regulator approves 5G spectrum auction rules, no Huawei ban," https://www.reuters.com/business/media-telecom/

brazil-regulator-approves-5g-spectrum-auction-rules-no-huawei-ban-2021-02-26/, https://www.cullen-international.com/news/2021/06/Most-LATAM-countries-have--use-or-lose-it--rules-to-promote-the-efficient-use-of-spectrum.html.

[7] Tomás, J. P., "France completes 5G spectrum auction," *RCR Wireless*, October 2, 2020, https://www.rcrwireless.com/20201002/5g/france-completes-5g-spectrum-auction.

[8] FCC, "DISH Network Corporation Petition for Waiver of Sections 27.5(j) and 27.53(h)(2)(ii) and Request for Extension of Time," https://www.fcc.gov/document/aws-4-dish-network-corporation-waiver-memorandum-opinion-and-order.

[9] Alleven, M., "Dish's 5G buildout revolves around wholesale vs. retail—report," *Fierce Wireless*, October 1, 2020, https://www.fiercewireless.com/wireless/dish-s-5g-buildout-revolves-around-wholesale-vs-retail-report.

[10] FCC, "3.5 GHz Band Overview," https://www.fcc.gov/35-ghz-band-overview.

[11] Lowe, J., "Why CBRS frequency bands are perfect for private LTE and private 5G networks," *Infovista*, https://www.infovista.com/blog/cbrs-frequency#.

[12] FCC, "3.5 GHz Band Overview," https://www.fcc.gov/35-ghz-band-overview.

[13] Alleven, M., "Google, CommScope, Federated and Sony get SAS certified for CBRS," *Fierce Wireless*, January 28, 2020, https://www.fiercewireless.com/wireless/google-commscope-federated-and-sony-get-sas-certified.

[14] FCC, "600 MHz Band," https://www.fcc.gov/wireless/bureau-divisions/broadband-division/600-mhz-band.

[15] Symons, H., and Milgrom, P., "Lessons from the US Incentive Auction," Inter Media, https://www.iicom.org/wp-content/uploads/im-jan2018-lessons-1.pdf; FCC, "How It Works: The Incentive Auction Explained," https://www.fcc.gov/about-fcc/fcc-initiatives/incentive-auctions/how-it-works.

[16] Augustino, S., "FCC Announces the Results of the $19.8 Billion Broadcast Incentive Auction," https://www.commlawmonitor.com/2017/04/articles/internet/fcc-announces-the-results-of-the-19-8-billion-broadcast-incentive-auction/; FCC, "600 MHz Band," , https://www.fcc.gov/wireless/bureau-divisions/broadband-division/600-mhz-band.

[17] Henry, C., "SES allies with Intelsat, Intel on revised US C-band proposal," *Space News*, February 18, 2018, https://spacenews.com/ses-allies-with-intelsat-intel-on-revised-us-c-band-proposal/.

[18] Arlen, G., "C-Band Auction Would Pay $9.7B starting in 2021," *Multichannel News*, February 10, 2020, https://www.nexttv.com/news/c-band-auction-would-pay-9-7b-starting-in-2021.

[19] ITU, "Agenda and Relevant Resolutions WRC-19," accessed July 24, 2021, http://handle.itu.int/11.1002/pub/80f52158-en.

[20] GSMA, "The WRC series: IMT Spectrum Between 24.25 and 86 GHz GSMA Public Policy Position," https://www.gsma.com/spectrum/wp-content/uploads/2018/12/AI-1.13-Positions.pdf.

[21] GSMA, "26 GHz and 28 GHz are both needed for 5G," https://www.gsma.com/spectrum/resources/26-ghz-28-ghz.

[22] FCC, "Use of Spectrum Bands Above 24 GHz For Mobile Radio Services, et al.," *Fifth Report and Order*, 34 FCC Rcd 2556 (3), (2019).

[23] FCC, "Use of Spectrum Bands Above 24 GHz For Mobile Radio Services, et al.," *Report and Order and Further Notice of Proposed Rulemaking*, 31 FCC Rcd 8014, 8165–67 (2016).

[24] Anritrsu, "FirstNet," https://www.anritsu.com/en-us/test-measurement/technologies/firstnet.

[25] Dano, M., "FirstNet: All AT&T LTE bands to be available to public safety this year" https://www.fiercewireless.com/wireless/firstnet-all-at-t-lte-bands-to-be-available-to-public-safety-year.

[26] Inside Towers, "Expanding 4.9 GHz Spectrum Use Will Take Time," October 2, 2020, https://insidetowers.com/cell-tower-news-expanding-4-9-ghz-spectrum-use-will-take-time/.

[27] Hardesty, L., "FCC says 'not so fast' on plan to lease 4.9 GHz to third parties," *Fierce Wireless*, June 1, 2021, https://www.fiercewireless.com/regulatory/fcc-says-not-so-fast-plan-to-share-4-9-ghz-third-parties

[28] Railway Technology, "India allots 5MHz spectrum in 700MHz band to railways," https://www.railway-technology.com/news/india-allots-spectrum-railways.

[29] Tarnutzer, B., "The 5G challenge for regulators – examining the cost of reserving spectrum for private networks," *GSM*, https://www.gsma.com/spectrum/examining-the-cost-of-reserving-spectrum-for-private-networks.

[30] Blackman, J., "Finland issues first private LTE/5G licenses, urges industry to pile in," *Enterprise IoT Insights*, July 7, 2020, https://enterpriseiotinsights.com/20200707/channels/news/finland-issues-first-private-lte-5g-licence-urges-industry-to-pile-in.

[31] 5G Networks, "Private 5G, 4G and Shared Spectrum," https://www.5g-networks.net/5g-technology/private-5g-4g-and-shared-spectrum/.

[32] Kostk, "White House announces reallocation of 3.45-3.55 GHz Band for 5G," *Red River Radio Amateurs*, August 10, 2020, https://rrra.org/post/2020/08/10/white-house-announces-reallocation-of-3.45-3.55-ghz-band-for-5g/.

[33] Hill, K., "FCC will vote on opening up 3.45-3.55 GHz for mobile networks," *RCR Wireless News*, September 9, 2020, https://www.

rcrwireless.com/20200909/spectrum/fcc-will-vote-on-opening-up-3-45-3-55-ghz-for-mobile-networks.

[34] Marcus, M. J., "Spectrum Policy for National Government Users: A Worldwide Policy Challenge," GTPRN, https://gtprn.org/2020/08/06/spectrum-policy-for-national-government-users-a-worldwide-policy-challenge.

[35] See also, ITU, "Spectrum pricing and trading," June 10, 2020, https://digitalregulation.org/spectrum-pricing-and-trading.

[36] Tomás, J. P., "European Commission to harmonize 26 GHz band for 5G deployments," *RCR Wireless*, https://www.rcrwireless.com/20190515/5g/european-commission-harmonize-26ghz-band-5g-deployments.

[37] Marti, M., "Which countries have met the EU's 5G deadline?" *PolicyTracker*, https://www.policytracker.com/blog/who-has-met-eu-5g-deadline-a-mobile-industry-darling-and-its-bete-noire.

[38] Shi, W., "Spain to auction 700 MHz for 5G, with sweetened conditions," *Telecoms.com*, June 01, 2021, https://telecoms.com/509944/spain-to-auction-700-mhz-for-5g-with-sweetened-conditions.

[39] 5G Americas Whitepaper, "5G Spectrum Vision," February 2019, https://www.5gamericas.org/wp-content/uploads/2019/07/5G_Americas_5G_Spectrum_Vision_Whitepaper-1.pdf.

[40] ITU, "GMPCS-MoU," https://www.itu.int/en/gmpcs/Pages/default.aspx. Such a regime generally exists already for MSS operations.

[41] ORBIT Act, U.S. Code 47 (2000), § 765f.

[42] BN Americas, "Brazil abandons plan to hold 5G auction this year," https://www.bnamericas.com/en/news/brazil-abandons-plan-to-hold-5g-auction-this-year.; https://developingtelecoms.com/telecom-business/telecom-regulation/11427-brazil-s-5g-spectrum-auction-will-it-now-happen-in-august.html; https://www.rcrwireless.com/20201125/5g/brazil-expects-5g -auction-take-place-end-h1-2021-report.

[43] Alleven, M., "SpaceX among those getting OK for new NGSO constellation," https://www.fiercewireless.com/wireless/spacex-among-those-getting-ok-for-new-ngso-constellation.

8

6G and Beyond

8.1 Introduction: 6G Status and Challenges

Based on recent precedent, almost every decade there is a new next generation wireless network technology developed and deployed for wide-spread commercial use [1]. Accordingly, although 5G is at early stages of deployment in some countries, and in early planning stages in others, the next generation of wireless networks, 6G is already being planned and technology development is underway [2]. 6G is being billed as the Internet of Senses, with hyperconnectivity expected between humans and everything [3]. Some wireless companies (terrestrial and nonterrestrial) are already working to start trials in the near term, while others have started to obtain patents for 6G technology. Similarly, standards groups are already developing standards for 6G deployment, while equipment vendors are looking to early 2030 as the time frame to have the first 6G equipment in the marketplace [3]. If this development continues on pace, using past history as a guide, we can expect to see 6G become a reality as early as 2030 [4]. This means that spectrum managers must be ready to address the challenges brought by 6G which will call for greater capacity and higher speeds to support the advanced applications envisioned. This growing need for access to spectrum has been the case for each generation of wireless technology and will continue for 6G (and beyond) as the world continues to increase its reliance on wireless technologies.

6G is expected to be an enhancement of 5G. It will continue to support all the myriad of use cases for 5G (e.g., augmented and virtual reality, anywhere, anytime super-low-latency communications), but will demand access to even more bandwidth to provide higher speeds and greater capacity. This means that access to additional frequency bands than those currently identified for use for 5G will need to be made available. This will require the increase of frequency bands through modified sharing or an increased efficiency mechanism along with availability of additional spectrum. Due to the increased capacity requirements it may be necessary to extend into the terabyte range to meet the myriad uses that 6G is claiming to be able to support. However, because of the wide-ranging use cases planned for 6G, regulators need to turn to the three-dimensional spectrum model discussed in Chapter 6. Regulators will also need to adhere to the principle of technology neutrality in making spectrum decisions. In addition, to meet user's capacity and security demands, we can expect 6G to begin to leverage nonspectrum-based wireless communications technologies, such as optical, which we are beginning to see early stages of deployment. Further in the future we may see other technologies such as quantum communications technologies deployed.

In this chapter we examine what 6G may look like, the challenges that spectrum managers are likely to face, and how to address these challenges. We also begin to look at the use of nonspectrum-based wireless communications technologies, and the role they may play in 6G and beyond wireless networks.

8.2 What Does 6G Look Like and What Is the Likely Impact on Spectrum Management?

To determine the spectrum requirements of 6G we need to first consider the technologies that will make up 5G and the uses cases for 6G. Of course, 6G will have to meet day-to-day requirements that we have today for voice and data technology (texting, e-mailing, web-browsing, and streaming) that have been provided with 3G, 4G, and 5G. In addition, 6G will most likely include the technologies not mature enough to be part of the 5G ecosystem. It is likely that any 5G technologies that did not make it into the 3GPP standards process by the early 2020s potentially will be a part of the 6G ecosystem.

Further, as with current and previous generations of wireless technologies, governments will want to enable the use of 6G to meet certain societal goals [5]. These include supporting the digital transition: the need for inclusion, addressing climate change and sustainability goals, furthering safety, and resilience. There will be increased pressure for 6G networks to provide wide-coverage with performance standards above 5G, both of which demand greater access to the spectrum resource.

Today, designers of 6G technology are looking to introduce new human-machine interfaces that extend the user experience across multiple physical and virtual platforms. These might include sensing, and immersive mixed realities providing the ultimate multi-media experience for a variety of use cases [6]. To meet these requirements, the 6G network will need to access the use of large bandwidths in existing and new spectrum bands. In addition to ensuring true global coverage and access, there is need to focus, really for the first time, on advancing the enablement of seamless multi-access service continuity. This service continuity will require using terrestrial and nonterrestrial networks to provide coverage across land, sea, and sky, efficiently addressing traffic and connection density. For 6G to be successful, perhaps even more than 5G, there willbe a need to harmonize spectrum management standards and regulatory structures across the globe and apply the use of a three-dimensional spectrum management model.

Based on the needs of 6G, we can expect the 6G network to have certain technical characteristics including:

1. Increased density of the network infrastructure for the large number of users and connected devices. The large number of access points will cooperate to form a cell-free network to provide a seamless quality of service over the coverage area [7].
2. Making intensive use of distributed processing and cache memories such as cloud RAN technology [8].
3. Meeting demands for wide-coverage of the 6G network using existing spectrum especially in the low- and mid-bands, as well as for additional spectrum use in the millimeter and beyond millimeter wave bands to provide the required high-capacity point to point links [9].
4. Requiring access to additional unburdened spectrum. As use of higher frequency bands is enabled, we can expect to hear calls for access to the THz bands by 6G proponents. This will depend on the ability of technology to operate successfully and be cost-effective in these bands.
5. Leveraging network slicing and multi-access edge computing to enable the development of new services with specialized performance requirements [10].
6. An increasing need for integration of terrestrial and nonterrestrial (e.g., satellite and drones) wireless networks, which can fill coverage needs (as well as resiliency requirements) and off-load traffic when there are heavy network loads [10]. Ensuring that nonterrestrial networks have access to sufficient spectrum to meet user demands will have to play a larger role in regulators' decisions on spectrum access.
7. Leveraging machine learning methodologies to improve spectrum resource allocation, assignment, and use to quickly meet expected very heavy spectrum needs [11]. Today regulators are only starting to use such technologies, but 6G will require a greater need for reliance on

dynamic spectrum access technologies to ensure sufficient bandwidth for all uses.

As 5G technologies morph into 6G, an increase in demand for spectrum is highly likely. For example, current AR technology requires 55.3 megabits per second (Mbps) to support 8K display (with one million points), which is viewed as being able to provide enough capacity for an adequate user experience on a mobile display [12]. However, to provide the truly immersive AR envisioned in the 6G world, the density needs to be significantly improved. This will require approximately 0.44 gigabits per second (Gbps) throughput (with 16 million points) to support things such as digital twins [13]. In addition, XR media streaming may have similar demands to 16K UHD (Ultra High Definition) quality video [14]. For example, 16K VR requires 0.9 Gbps throughput (with compression ratio of 1/400) [13]. Accordingly, the current user experienced data rate of 5G is not sufficient for seamless streaming and additional capacity to support these needs will be required.

The same challenges exist with holograms that are anticipated to be an important of the 6G experience. To provide hologram display as a part of real-time wireless services, extremely high data rate transmissions, hundreds of times greater than the current 5G system, must be met [14]. Moreover, the support of some much-anticipated holograms using a large number of pixels require increased spectrum. For instance, a human-sized hologram requires several Tbps of spectrum. With the current peak data rate of 5G at 20 Gbps, clearly 5G will not be able to support the extremely large volume of data required for hologram media in real-time. We expect other 6G requirements will put additional strains on the spectrum resource.

The many new services of the 6G era will require a huge amount real-time data processing, a hyper-fast data rate, and extremely low latency. By leveraging these characteristics, these new services will be able to support the envisioned use cases through hyper-connectivity. However, these new services will require access to vast amounts of additional spectrum capacity by terrestrial and nonterrestrial wireless technologies. To realize advanced multimedia services such as truly digital replica, 6G must provide a much higher data rate than 5G. 5G is designed to achieve 20 Gbps at peak data rate. In 6G, the goal is to provide the peak data rate of 1,000 Gbps and a user experienced data rate of 1 Gbps [14].

In addition to speed, service networks will need to have wide coverage areas and sufficient capacity to support the number of expected users and devices. 6G is envisioned to have a large density of users due to the hyper-connectivity with everyone and everything connected. In addition, there needs to be greater capacity, including in rural areas, as a larger number of devices are expected to be supported to meet expected user experience demands. For example, some estimate that the number of connected machines will require 6G to support about ten million devices per square kilometer [15]. This is ten times larger than the connection density requirement of 5G [16]. In addition, in the 6G era, users will expect seamless high-end services indoors and outdoors to support their everyday lives [17].

8.3 Regulators and Spectrum Managers Must Begin Planning to Meet 6G Spectrum Needs

Understanding the use cases discussed above helps us better understand the challenges that spectrum managers will face with 6G (and beyond). Besides the increased need for speed, capacity, and wide-area coverage, in building penetration of service will also be key. To achieve additional spectrum availability in the conventional low-, mid-, and high-frequency bands will be critical, but 6G will require far more.

We are still very early in understanding the full spectrum management challenges of 6G. We expect to see many of the same challenges we face today including the hesitancy or even unwillingness of incumbents to share or vacate spectrum to enable more efficient uses. There likely will be a need for additional spectrum that can only be handled by using extremely high-frequency bands which have only been used sparingly to date. In addition, it can be expected that these tensions may increase as the 6G network includes greater reliance on nonterrestrial networks.

Below we discuss the continued challenges regulators need to address to enable spectrum sharing, to make the most efficient use of the spectrum resource, and to make THz frequency bands available for future use.

8.3.1 Technology Neutral Decisions

Although technology-neutral decision making in spectrum management has often been a stated goal, quite honestly, it has rarely been implemented. This is demonstrated by our discussions of recent spectrum decisions regarding the 28 GHz band, CBRS, and the 3.5 GHz band. Those cases involved making additional spectrum available for mobile wireless services at the expense of satellite use. In the 5G world, where terrestrial wireless serves as the basis for most use cases with satellite simply a complimentary technology, this may make sense. However, in the 6G world where it is likely that nonterrestrial technologies (satellite, HAPs and drones) are likely to play a larger role than in 5G, regulators must ensure their decisions are technology neutral, a difficult task.

Regulators will need to more carefully balance the outcome of their decisions to allow growth of multiple technologies. The result can no longer growth limited to terrestrial wireless technologies. For example, a better approach for enabling 6G in the 28 GHz band [18] (as opposed to limiting satellite use to the most rural and remote areas) could involve greater flexibility through coordination or the use of dynamic access technologies while still providing terrestrial wireless with access to sufficient bandwidth to meet their user needs. This would result in a more balanced approach and, consistent with the goal of spectrum managers, a more efficient use of the 28 GHz band. Accordingly, as 6G becomes a reality, it is going to be critical for spectrum managers to move away from their preference for terrestrial wireless technologies looking to enable greater use of the spectrum resource by terrestrial and nonterrestrial technologies.

8.3.2 The Need for Use of a Three-Dimensional Spectrum Model

Along with technology-neutral decision making is the need to develop a three-dimensional spectrum model. While 5G begins to call for use of such a model, the 6G world demands such use. In the 6G *Internet of Senses* world, wireless communications will be occurring on the ground, on street poles, on drones or balloons, on HAPS, on satellites in different orbits, in spacecraft, on the moon, and beyond. Accordingly, the interference environment will be drastically modified from our current ground-based with towers usage that has been the driving focus of today's spectrum management.

The three-dimensional spectrum model will need to look at spectrum use to determine the technical protections needed to guard against interference. While, in some cases it may limit spectrum use, in others, it may increase or make more efficient use of the spectrum if multiple technologies share the same frequency bands. It will be important for spectrum managers to look at the spectrum resource in three dimensions to gain the greatest use of the resource while minimizing interference.

8.3.3 Continued Challenges in Sharing and Making the Most Efficient Use of the Spectrum Resource

Even today, regulators must consider how to increase sharing and make more effective use of the existing spectrum bands. While, as we saw in Chapters 6 and 7, regulators are actively working towards new and more efficient spectrum uses, this effort is still in its early phases. To design 6G networks that are responsive to the diverse needs of users, with appropriate speeds, coverage, and capacity will require access to more spectrum in all existing bands for a variety of terrestrial and nonterrestrial technologies. 5G is the first generation of wireless technology being seen as a network of different technologies. However, unlike 5G technologies, the 6G world requires the integration of both terrestrial and nonterrestrial spectrum, and enhanced technologies to deliver a hyper-connected world.

As regulators open new bands to sharing, they must be careful to ensure protection for incumbent services that will continue to be used in the future, including 6G. A good example is the ongoing fight over the 6 GHz band for use by Wi-Fi versus IMT (or 5G mobile) in a band with incumbent satellite services. In determining the ultimate use of the band by mobile wireless (whether unlicensed of licensed IMT), regulators must look carefully at the impact on FSS and MSS operators with feeder links in the band to ensure protection of incumbent services. While these may not be part of the 5G eco-system today that is likely to change as 6G begins to depend more heavily on space-based systems. Failure to look at the long-term use requirements, as opposed to those in the short term (i.e., for 5G) may hamper the ability of 6G and beyond technologies in the future.

Spectrum managers should also consider the advent of new technologies that can be leveraged for increased spectrum sharing. The on-going development of new technologies will allow for greater flexibility in sharing capabilities. For example, we can expect advances in spectrum-access decision making to enable sharing through AI-inspired algorithms to increase spectrum sharing by giving real -time access to spectrum. These technologies may increase flexibility for spectrum use when 6G and beyond becomes available.

Examining interference issues, such as aggregate interference too late in the process can have serious ramifications for spectrum use. Where sensors are critical, discovering the impact of aggregate interference late in deployment of technologies could force the transition of co and adjacent frequency services out of the frequency band affecting economic ROI and user access to services. Accordingly, regulators must begin to consider the impacts of aggregate interference in making spectrum decisions.

8.3.4 The THz Frequency Band

Even in a world where spectrum sharing continues to expand into more frequency band, 6G is going to require access to additional frequency bands. For this reason there has to be a serious examination of the use of the THz frequency bands by service providers, regulators, and the ITU [19]. The THz band is made up of an enormous volume of available bandwidth, which will enable the use of the extremely wideband channels with tens of GHz-wide bandwidth that 6G requires [20].

However, with 5G networks being extremely localized due to the inherent limits of millimeter waves, the same problem may occur when we deploy 6G networks and start to utilize the THz bands. For example, the range of terahertz radiation is around 10 meters, which is much too short for significant 6G coverage, with potential for significant path loss [21]. In addition, the effect of atmospheric absorption (i.e., absorption by molecules in air) in the THz band is more severe than in lower frequencies, as the absorption lines for oxygen and water are mostly located in the THz band. This will require further development of tractable THz multipath channel models for indoor and outdoor environments [22]. To adequately utilize the THz band, work needs to be done to extend the signals for the THz frequency bands or the utilization of these bands for 6G will remain severely limited.

Despite these challenges, regulators should (some already are) start to look at ways to make access to the THz frequency bands available sooner. For example, in March 2019, the FCC opened the spectrum between 95 GHz and 3,000 GHz for experimental us,e and unlicensed applications to encourage the development of new wireless communication technologies [23]. In addition, there are increasing discussions on use cases and deployment scenarios for 5G new radio (NR) systems operating at bands beyond 52.6 GHz in a variety of forum [24]. In addition, on November 6, 2020, China successfully launched an experimental test satellite with candidates for 6G technology into orbit, along with 12 other satellites, using a Long March 6 launch vehicle rocket. The satellite's stated purpose is to verify the terahertz (THz) communication technology in space [25]. As 6G development progresses with technologies that support bandwidth intensive use cases, we can anticipate 6G designers will find ways to utilize the THz bands in ways that make sense.

Despite a myriad of challenges, use of the THz frequency bands brings other benefits likely to help drive development for many reasons [26]. For example, communications in the THz bands can potentially provide high-precision positioning capability by enabling accurate ranging between transmitter and receiver. In addition, THz bands are more secure than lower frequency bands, a critical need for the 6G network. Further, THz frequency bands offer very high speeds, and very low latencies; characteristics that are key to meet the anticipated 6G use cases. THz communication will contribute to enable interesting, needed applications such as immersive virtual reality (e.g., remote medicine), ultra-broadband connectivity for rural areas (e.g., where fiber is not possible), and even a terabit backhaul in the sky (above the atmosphere, where there is no absorption). Also, by leveraging the very small size of THz transceivers and antennas (sub-millimetric), THz communication may enable the development of transformative applications such as nano-biosensing wearable and implantable devices, wireless brain machine interfaces that can interact with neurons, and more yet to be developed. With so many possible benefits it seems likely that technologists will work towards finding a way to best access and utilize the THz frequency bands.

From a spectrum management point of view, we can expect to continue to see experimental uses of these frequency bands, which regulators should work to enable with minimal burdens. In addition, regulators must work to make these bands available for 6G uses for all communications technologies, on the ground or in the air. This requires appropriate efforts on both a global and national level. Globally, the ITU will need to allocate these bands to appropriate radio services to help harmonize their use. Nationally, countries will need to then implement these allocations domestically. As noted earlier, the spectrum allocation process takes significant amounts of time. With the WRC-23 Agenda already finalized with no mention of THz spectrum planned for discussion, it will be critical for regulators to focus on the WRC-27 Agenda to make sure that the ITU takes appropriate actions on a timely basis so these changes can be implemented globally. Failure to take actions in the near term to make sure there is sufficient spectrum for 6G will leave the world unprepared, or result in spectrum fragmentation as more advanced nations moving toward 6G deployment take individual actions. This may be a good time to revisit the frequency allocations of these bands, which were done in a very different time-period, to see if previous approaches still make sense.

Finally, there must be consideration of how to assign licenses and regulate these frequency bands. Since they are lightly used, we have an opportunity as we look at their assignment and use to find a way to make more efficient use of the THz radio spectrum than we have for previous bands. This might be the appropriate band to provide more flexibility and increased sharing requirements from the start, such as those utilizing artificial intelligence and dynamic access tools. In addition, we should seek to avoid imposing unnecessary regulatory burdens. Taken together, this may allow for a different approach to assignment than used in the past. Regulators should get ahead of this effort and start to look for novel ways to assign frequencies today. If they wait until use is required it will likely hamstring the use of existing approaches that do not take into account the fallow nature of many of these bands. By beginning to plan for use of the THz frequency bands now, regulators can ultimately strike the appropriate balance for increased efficiency in the use of this valuable spectrum resource.

8.3.5 Greater Accountability for Use of Spectrum

One area where spectrum managers continue to underperform is in ensuring that the spectrum allocated and assigned actually is put to use. There are a myriad of papers and analyses demonstrating that there is a significant amount of spectrum that has been assigned to licensees that is simply not being utilized. While spectrum managers have used build-out and other requirements to ensure spectrum use, more often than not, they fail to achieve their goals either through lack of enforcement, through time extensions to build-out, or general forgiveness of build-out requirements.

In a 6G world where increasing capacity will be required spectrum managers will need to act to increase spectrum use accountability. At a minimum greater enforcement needs to be applied where build-out is not accomplished. However, spectrum managers will need to look for ways to leverage technologies ensuring spectrum is being used efficiently to increase spectrum use. It is imperative that spectrum regulators address the very real issue of assigned spectrum laying fallow if there is going to be sufficient access to spectrum to support 6G and beyond requirements.

8.4 Nonspectrum Solutions to Connectivity: Optical and Quantum

Given the extreme performance demands of the planned 6G use cases, it is likely that there will be a need to use innovative technologies to help meet these requirements including having greater speed and capacity along with increased security. Two of these burgeoning technologies are optical wireless networks and quantum communications technologies. Optical wireless is already being incorporated into some terrestrial and nonterrestrial networks, including for terrestrial wireless back-haul and satellite intersatellite links [27]. However, it is still early to determine how much traffic and tthe role optical wireless technologies will play. However, regulators and policymakers should certainly pay attention to this.

Quantum communications development is even farther behind optical wireless. However, there have been some early tests, including some by China using a satellite network, that have been successful. While it is doubtful that quantum communications will be ready for use for 6G, in the future this technology will likely play an important role, if successfully commercialized.

8.5 Conclusion

6G brings great promise and challenges as there is a movement to utilize higher frequency bands along with the need for additional sharing scenarios. By starting to plan now, regulators can get ahead of any potential challenges. As the THz bands are explored and new sharing scenarios are introduced, in new or in existing frequency bands, regulators must look to ways that provide operators with the flexibility they need without imposing undue burdens. We should not see a return to today's regulations but should see the creation of regulations which makes the most possible efficient use of the spectrum by enabling flexibility and technology neutrality.

References

[1] Rabaey, J., "What is 5G? The next generation of wireless, explained," *The Conversation*, https://theconversation.com/what-is-5g-the-next-generation-of-wireless-explained-96165.

[2] Mishra, Y., "6G Progress Report: China owns 35% of the overall patents and Huawei to release two satellites in July," *Huawei Central*, April 28, 2021, https://www.huaweicentral.com/6g-progress-report-china-owns-35-of-the-overall-patents-and-huawei-to-release-two-satellites-in-july/; Szporer, R., "Navigating 6G Spectrum & Standardisation Roadmaps Requires Coordination," *6G World*, May 6, 2021, https://www.6gworld.com/

exclusives/navigating-6g-spectrum-standardisation-roadmaps-requires-co-ordination.

[3] GreyB, "Towards 6G network: Top companies and universities mapping the 6G technology," https://www.greyb.com/6g-companies/.

[4] Mint, "Move over 5G, Japan plans to launch 6G by 2030, says report," https://www.livemint.com/technology/tech-news/move-over-5g-japan-plans-to-launch-6g-by-2030-says-report-11579590569894.html.

[5] Mint, "Forget 5G, the US and China are already fighting for 6G dominance," https://www.livemint.com/news/world/the-6g-fight-heats-up-between-us-and-china-11612925167992.html.; Nikkei Asia, "US and Japan to invest $4.5bn in next-gen 6G race with China," https://asia.nikkei.com/Business/Telecommunication/US-and-Japan-to-invest-4.5bn-in-next-gen-6G-race-with-China.

[6] GreyB, "Towards 6G network: Top companies and universities mapping the 6G technology," https://www.greyb.com/6g-companies/.

[7] Akhtar, M. W., Hassan, S. A., Ghaffar, R. et al., "The shift to 6G communications: vision and requirements." *Hum. Cent. Comput. Inf. Sci.* 10, 53 (2020). https://doi.org/10.1186/s13673-020-00258-2.; Bhat, J.R., AlQahtani, S., "6G Ecosystem: Current Status and Future Perspective," IEEE Access, January 26, 2021, https://ieeexplore.ieee.org/stamp/stamp.jsp?arnumber=9335927.

[8] Raddo, T. R., Rommel, S., Cimoli, B. et al., "Transition technologies towards 6G networks." *J Wireless Com Network* 2021, 100 (2021), https://doi.org/10.1186/s13638-021-01973-9.; Parsons, E., "The four key components of Cloud RAN," *Ericsson*, https://www.ericsson.com/en/blog/2020/8/the-four-components-of-cloud-ran.

[9] Akhtar, M. W., Hassan, S. A., Ghaffar, R. et al., "The shift to 6G communications: vision and requirements," *Hum. Cent. Comput. Inf. Sci.* 10, 53 (2020), https://doi.org/10.1186/s13673-020-00258-2.

[10] NITRD, "Artificial Intelligence & Wireless Spectrum: Opportunities and Challenges, 2019 Workshop Report." (November 2020), https://www.nitrd.gov/pubs/AI-WirelessSpectrum-2019WorkshopReport.pdf.

[11] Eswar, "Samsung says 6G in 2028 with peak speeds of 1,000Gbps.," *TechNuws*, July 14, 2020, https://technuws.com/samsung-says-6g-in-2028-with-peak-speeds-of-1000gbps/.

[12] Carlyon, C., "6G what, when and why, cows that don't burp and dam robots," *Substack*, https://whatsoncraigsmind.substack.com/p/6g-what-when-and-why-cows-that-dont.

[13] Eswar, "Samsung says 6G in 2028 with peak speeds of 1,000Gbps.," *TechNuws*, July 14, 2020, https://technuws.com/samsung-says-6g-in-2028-with-peak-speeds-of-1000gbps/.

[14] Tabari, M., "Samsung's bizarre 6G vision for 2028 includes holograms, digital replicas and extended reality," *Laptop Mag*, https://www.laptopmag.

com/news/samsungs-bizarre-6g-vision-for-2028-includes-holograms-digital-replicas-and-extended-reality.

[15] Tonex, "Introduction to 6G I IMT-2030," https://www.tonex.com/training-courses/introduction-to-6g-imt-2030.

[16] Giordani, M., et al., "6G: Towards a Fully Digital and Connected World," *6G Wireless Summit*, Levi, Finland, March 26th, 2019, http://www.6gsummit.com/wp-content/uploads/2019/04/Day3_Session7_Giordani_PadovaUniversity.pdf.

[17] Giordani, M., et al., "6G: Towards a Fully Digital and Connected World," *6G Wireless Summit*, Levi, Finland, March 26th, 2019, http://www.6gsummit.com/wp-content/uploads/2019/04/Day3_Session7_Giordani_PadovaUniversity.pdf.

[18] GSMA, "The WRC series: IMT Spectrum Between 24.25 and 86 GHz GSMA Public Policy Position," https://www.gsma.com/spectrum/wp-content/uploads/2018/12/AI-1.13-Positions.pdf.

[19] ITU, "Technology trends of active services in the frequency range 275-3 000 GHz," Rep. ITU-R SM.2352-0 (2015), https://www.itu.int/dms_pub/itu-r/opb/rep/R-REP-SM.2352-2015-PDF-E.pdf.

[20] Stack, J., "6G Proof-of-Concept System Is Showcased by Samsung and UCSB," *Engineering.com*, July 4, 2021, https://www.engineering.com/story/6g-proof-of-concept-system-is-showcased-by-samsung-and-ucsb.

[21] RP-Photonics.com, "Terahertz Radiation," July 26, 2021, https://www.rp-photonics.com/terahertz_radiation.html.

[22] Sarieddeen, H., et al., "An Overview of Signal Processing Techniques for Terahertz Communications," *IEEE*, https://arxiv.org/pdf/2005.13176.pdf.

[23] Alleven,M.,"FCCmovestoopenspectrumabove95GHzfornewtechnologies," https://www.fiercewireless.com/wireless/fcc-moves-to-open-spectrum-above-95-ghz-for-new-technologies.

[24] Tervo, O., et al., "5G New Radio Evolution Towards Sub-THz Communications," https://arxiv.org/pdf/2002.02659.pdf.

[25] BBC News, "China sends 'world's first 6G' test satellite into orbit," 2021, https://www.bbc.com/news/av/world-asia-china-54852131.

[26] RF Wireless, "Advantages of 6G wireless/disadvantages of 6G wireless," https://www.rfwireless-world.com/Terminology/Advantages-and-Disadvantages-of-6G-Wireless-Technology.html.

[27] Internet Society, "Solutions for Traffic Backhaul in Community Networks," https://www.internetsociety.org/resources/doc/2019/solutions-for-traffic-backhaul-in-community-networks; "Space—Twenty Years of ESA Effort," ESA,https://www.esa.int/esapub/bulletin/bullet91/b91lutz.htm.

9

Conclusion: Challenges Ahead

9.1 Introduction

We are potentially at a momentous time in world history if 5G and beyond technologies realize their full potential. However, even if these technologies only deliver half of what they promise, one thing is clear: access to additional spectrum will be required to meet the growing demands for 5G terrestrial and nonterrestrial wireless services. Along with the need for additional spectrum access come the many challenges discussed in this book. To enable the introduction of innovative technologies, and for us to be successful at making the most efficient use of the radio communications spectrum, these challenges must be addressed by spectrum managers.

We will examine five major challenges and propose possible solutions to fulfill spectrum access for 5G and beyond demands. As discussed below, the solutions have pros and cons, but it is important for spectrum managers as the demand for spectrum to support new generations of wireless networks increases, to at least explore these and other possible solutions.

9.2 Greater Spectrum Use

Increasing spectrum use is always a critical goal of spectrum managers and one that is very difficult to define and attain. While the easiest way to determine if spectrum is being used appears to be monitoring it's use, a survey of the literature shows no spectrum manager or regulator that requires proof that a radio spectrum licensee is efficiently using their spectrum assignments. Spectrum

Table 9.1
Six Major Challenges Ahead and Possible Solutions

Challenge	Example of Possible Solutions
Increasing the Use of the Spectrum Resource	Imposing more stringent build out requirements, including monitoring potential use, and, as a last resort, explore increased spectrum sharing. This will ensure that there is greater use of the spectrum resource.
Ensuring the Protection of Incumbent Uses	Requires spectrum managers to early on fully understand the impact of changes to incumbent uses and how to best address them through, strategies such as interference protection and/or reassignment possibly combined with funding to address interference concerns or to relocate operations.
Making Additional Spectrum Available on a Technology Neutral Basis	Spectrum managers must ensure there is a balanced approach to making spectrum available among technologies by having a fact-based rationale for their spectrum allocation and assignment decisions.
Reducing Regulatory Burdens and Fees	Ensuring that regulations are narrowly drawn and fees are cost-based and reviewing on at least a bi-annual basis regulations to ensure they are necessary.
Achieving Global Harmonization	Working with other regulators on a bi-lateral and multi-lateral basis on spectrum planning.
Incenting Competition	Enabling multiple competitors to have access to spectrum.

managers do impose a range of requirements including build-out requirements, notice requirements of spectrum use, and reporting on the number of subscribers to demonstrate spectrum use. However, there is little evidence of regulators actually requiring efficient use of the radio spectrum resource. As the literature shows, this means that there are likely many instances that radio spectrum licensees are not making efficient use of their spectrum [1, 2].

Many spectrum managers have tried to encourage greater spectrum use conditioning licenses on such things as notifying the regulator ofimplementing license use by a certain date, or imposing build-out requirements including coverage requirements. However, most of these requirements only address a small fraction of the potential licensed spectrum use. This raises three questions; what can be done to increase efficient use of the spectrum resource; what can be done if required spectrum use is not performed to remedy this situation; and what incentives can be given to licensees to make more efficient use of the spectrum resource.

One potential remedy is to increase the build-out requirements and milestones on radio spectrum licensees. A brief survey of build-out requirements shows that often build out is only required for a small percentage of the population or geographic area. Regulators could require increased build-out requirements to increase use, but there would be an economic cost to the licensee, as well as potential impact in auctioned frequency bands on the amount of money a licensee may be willing to spend for a particular service area. In fact, as we have seen in the past, auctions with too stringent conditions, can fail [3, 4].

It is even harder to determine the appropriate remedy if a licensee meets use or build-out requirements. Financial penalties would likely backfire, potentially taking away economic incentives to build out these areas. Another approach that countries such as Brazil have been considering is to take back spectrum from the licensee where build out is not achieved to have it relicensed. While this may be a solution, the new licensees may face similar economic challenges to build-out.

At the 2019 WRC an approach was agreed upon to increase the efficiency of spectrum resource use for NGSO satellite systems.

This was because traditionally the ITU has only required satellite operators to BIU their spectrum and orbital locations by having a satellite capable of communicating in the required spectrum at the orbital location for 90 days before their ITU rights expire. However, due to the desire of countries to ensure efficient use of the NGSO spectrum and orbit resource, at WRC 19 a more stringent approach was adopted for NGSO systems operating in popular FSS frequency bands like the Ku and Ka bands. Specifically, following their BIU, satellite network operators will have to deploy 10% of their constellation within 2 years after BIU, 50% within 5 years, and complete deployment within 7 years or face losing access to some of their spectrum resource. It is too early to see if such an approach works resulting in greater use of the spectrum resource for NGSO FSS satellite networks.

And finally, in terms of incentivizing greater use of the spectrum resource to date, we have seen few, if any approaches that have worked, although imposing large fees, such as through auctions for access to spectrum, provides a good financial incentive that is lacking using other means. For example, enabling secondary markets by spectrum licensees has not been widely successful for competitive or economic reasons.

So, what is the solution? It might be that spectrum managers need to seriously think about using spectrum sharing methodologies, such as White Spaces, CBRS, or 3.5 GHz, that enable alternate use of licensed spectrum where primary use is not being made. In the case of the 3.5 GHz, incentives are provided to operators to increase efficiency. Such an approach is extremely controversial and may be difficult to implement, but in frequency bands where regulators know that spectrum is not put into wide-spread use, it may be worthwhile to explore such approaches in future bands. Another approach may be similar to what the ITU did for NGSO systems; impose milestones with build-out requirements where some of the spectrum access is lost over time if not used. While not perfect, it may enable the opening of unused spectrum on a more timely basis for other uses.

One other possible solution to be considered is reducing guard bands that exist throughout the international and domestic tables of allocations where possible. With the improvement of technologies, it may be possible to increase access to the spectrum resource by carefully modeling the needs for guard bands using current technology. Since many of the guard bands in place today are based on modeling with antiquated technology, this may be an area where increased access to the spectrum can occur.

9.3 The Protection of Incumbent Uses

As spectrum regulators seek to make access to additional spectrum available for 5G and beyond uses, many of these bands will have incumbent uses. The challenge is to ensure that important incumbent uses are protected, in the same band or transitioned elsewhere so they can continue to operate successfully. This is vital since many incumbents provide important services including 5G, such as satellite networks in the millimeter wave bands. These incumbent services range from commercial satellite operations supporting broadcast communications to critical government services, such as military radars. Therefore, spectrum managers, when making decisions about changing spectrum use must be cognizant of the impact this could have on current incumbents.

For this reason that it is critical for spectrum managers understand the status quo of incumbent services, and the impact of disruption of this service. With this understanding spectrum managers are better able to protect the incumbent use and where appropriate, allow further growth before beginning proceedings to move the incumbents. As part of this analysis, the spectrum manager should also identify options available to protect the incumbent use along with the potential costs of moving the service to other bands. This analysis should be completed prior to starting to move services because of the importance of understanding the source of funding required to absorb the cost of transferring services, and what is acceptable to the incumbent operators.

If spectrum managers move without this full understanding, it will be difficult and take significant time, to successfully free up the use of spectrum for next generation uses. Working with incumbent users to understand their needs, including how to reimburse them for costs, will make spectrum managers more successful in their efforts to open additional spectrum for 5G and beyond.

In some cases, it may be in the public interest not to introduce new uses in the band. It is likely there will always be frequency bands where the right decision to protect important uses is to leave the status quo, at least for the interim. This is a hard decision for spectrum regulators to make, but one that must be considered.

9.4 Making Additional Spectrum Available on a Technology Neutral Basis

Although many spectrum managers and governments pay lip service to the policy of technology neutrality in spectrum management, few have been successful, in large part, due to the risk of harmful interference to others if usage is changed. For many countries, the primary economic pressure is that auction revenues are needed to pay for the operations of the regulator itself or support the government's treasury. In addition, regulators and politicians have an easier time supporting spectrum needs for technology utilized on a day-to-day basis, like mobile phones, than less prominent technologies. This is underscored by the large economic power of the mobile terrestrial wireless companies.

Nonetheless, it is critical to ensure that all technologies that will compromise 5G and beyond have access to spectrum in a balanced manner along with technologies, such as weather satellites, not directly part of 5G and beyond that provide critical wireless services. Unfortunately, as some recent examples demonstrate, spectrum managers have recently stepped away from following the principle of technology neutrality in spectrum management [5].

Accordingly, it is important that spectrum managers take a broader view of spectrum allocation and assignment than simply focusing on next generation wireless networks and more specifically, terrestrial spectrum, or adding monies to their countries' treasuries. A few steps can be taken to achieve neutrality in spectrum management. First, spectrum managers must have a realistic understanding of demand for spectrum services and utilization. Understanding current and future uses of a given frequency band

and the possible negative outcomes resulting from changing current assignments helps establish spectrum management neutrality. A good example of this was the FCC's Spectrum Frontiers proceeding, where the FCC opened the 24 GHz and the 28 GHz bands to terrestrial mobile wireless 5G services at the cost of weather satellites and commercial broadband satellite use [6]. At the time it was thought that limiting use of these and adjacent bands by these systems was a critical national interest to ensure that 5G reached everywhere to address the digital divide [6]. However, as FCC Chairwoman Rosenworcel noted if this spectrum was just allocated to 5G it would actually result in an increase in the digital divide [7].[1]

Spectrum managers need a realistic understanding of what spectrum incumbents and future users will need to be successful. Spectrum managers must strive to understand current users, and their demand for future growth. They must also work to obtain a realistic understanding of future demands by new uses, such as 5G terrestrial mobile, along with competing new uses, such as Wi-Fi 6. Having a factual, reasoned basis for making spectrum decisions may help avoid a lack of balance of spectrum allocations and assignments between the different services.

9.5 Reducing Regulatory Burdens and Fees

In recent years, some spectrum managers and regulators have appeared to move away from regulatory *light touch* non-cost-based regulatory fees. In fact, we have seen a return to heavy-handed regulation, if seemingly innocuous, like requiring reporting requirements to update certain service data without a compelling public interest reason, or having short-term licenses (e.g., one year long) requiring constant renewal by the licensee and the payment of fees. These requirements impose significant regulatory burdens and costs, without a corresponding benefit. In addition, there is a continuing trend by regulators to base regulatory fees on more than administrative costs, imposing non-cost-based fees on licens-

1. Despite this decision, acting FCC Chairwoman Rosenworcel recently noted that "if we just relied on millimeter wave spectrum, we would actually increase the digital divide for 5G."

ees, or considering the imposition of other regulatory measures, such as bonds, as the United States does for U.S. satellite systems, resulting in significant costs to licensees. These unnecessary additional costs are ultimately passed down to consumers, increasing the charges they must pay and, in some cases, increasing the digital divide. Further, such actions make the use of the spectrum resource less efficient by imposing unnecessary costs and burdens on users which acts to deter use of the resource.

One good example of an unnecessary regulation and imposition of fees is the lack of blanket license regimes for very small aperture terminals (VSATs) used by end-users to receive satellite broadband service. VSAT terminals that provide satellite broadband to end-users have virtually identical technical configurations, so to streamline regulation and reduce costs, some countries, (such as the United States and Canada) provide blanket licensing for VSAT terminals. This means a single license is issued with a single cost-based fee paid for the licensing of all VSATs of a certain model used by a satellite broadband provider in that country. Unfortunately, several countries, such as Ecuador, have failed to adopt such a regime, and instead are levying user fees of roughly $10 USD and larger for each VSAP used in-country. The rationale for such a regime is unclear. Since VSATs have the same interference parameters it is easy to license them, as mobile phones are today, in a single license with a single fee. It appears the major reason for the continuation of such an unnecessary regime is the ability to recoup fees. Unfortunately, this approach neglects the end-users who are the ones that suffer from such unnecessary regulation, negatively impacting the digital divide, and reducing use of the frequency bands where satellite broadband operates.

Another example is revenue-based fees. Governments are continuing to attach fees based on a percentage of revenues earned by spectrum licensees. While providing steady income to government coffers, these fees can incentivize operators to limit the business in the country to avoid the high fees. This may result in the licensee with such a fee requirement not making full use of the spectrum resource to avoid burdensome fees.

9.6 Achieving Global Harmonization

The importance of harmonization of the global spectrum resource increases with each generation of wireless technology to obtain the cost efficiencies of a widely deployed technology. Harmonization also improves other important services such as those used for radio, astronomy, and the weather. Harmonization can be experienced in different ways, including the impact on the reduction of cost in equipment and communication services. These are best achieved through economies of scale that global harmonization would provide. However, oftentimes global harmonization suffers at the hands of domestic regulators who fail to consider how domestic spectrum decisions can impact the global spectrum management process.

A good example of this discussed earlier was the use the United States made of some millimeter wave bands in its Spectrum Frontiers proceeding. For example, in the 24 GHz frequency band, the FCC adopted different protection criteria for adjacent bands where critical weather satellites operate globally to help monitor things such as climate change. In this situation, the United States approach was largely rejected globally at the ITU causing some negative consequences. First, the United States will need to do it alone with development of 5G mobile wireless technology in the 24 GHz band due to the different interference protection criteria it chose. Second, global weather systems are hampered in their goal for uniform measurements in the United States and the other countries that took the nonconforming approach.

It would be foolish to expect countries to always employ globally harmonized spectrum, but it is important that spectrum managers carefully consider the broader consequences of these decisions before taking them.

9.7 Competition

A final important consideration for spectrum managers and regulators is to incentivize competition. As decisions are made about spectrum use, it is important to keep in mind the impact of direct and indirect competition. Competition in this case is not just competition among the same service (e.g., among mobile wireless service providers), but among technologies. For example, low-latency, high speed NGSO satellite systems providing 5G broadband services may compete directly with high speed fixed wireless 5G broadband providers. In this case, when both entities require access to the spectrum resource the regulator should ensure that the allocation of the spectrum resource does not foreclose competition between services in order to incentivize competition.

In addition, we see an increasing trend toward government ownership of terrestrial and nonterrestrial wireless networks.[2] As seen in the past, government ownership of communications providers often results in benefits bestowed on these entities, including greater access to spectrum resource, faster processing of applications, as well as favoritism in government procurements, which results in a further increase in government funding. It is important that these governments, not limit competition by providing preferential treatment of these entities to access to spectrum over competitors.

Accordingly, spectrum managers and regulators must ensure that no one entity is able to foreclose competition in a wireless market through the assignment of spectrum. Doing so will likely result in higher costs for consumer and less innovation in the marketplace.

9.8 Conclusion

We are in a unique period with the opportunity to unleash a world of new technologies and spectrum uses that will provide great benefits to the world as 5G and 6G are globally deployed. However, governments, international organizations, and stakeholders must work jointly to ensure that we make the most efficient use of the spectrum resource. Despite the competitive battles over spectrum, spectrum managers must ensure that all technologies are included in each generation of wireless technology. By addressing the challenges we've discussed spectrum managers can success-

2. For example, recently we have seen Canada and the United Kingdom make significant investments in NGSO satellite networks, Telesat and OneWeb, respectively.

fully establish a spectrum regime that can achieve the goals of a 5G and beyond world bringing the benefits of these new technologies to everyone, wherever they are.

References

[1] Bleicher, A., "Peaceful Coexistence on the Radio Spectrum," *IEEE Spectrum,* https://spectrum.ieee.org/telecom/wireless/peaceful-coexistence-on-the-radio-spectrum.

[2] Szporer, R., "Navigating 6G Spectrum & Standardisation Roadmaps Requires Coordination," *6G World,* May 6, 2021, https://www.6gworld.com/exclusives/navigating-6g-spectrum-standardisation-roadmaps-requires-co-ordination/.

[3] Cave, M., and Nicholls, R., "The Use of Spectrum Auctions to Attain Multiple Objectives: Policy Implications," *Science Direct,* Volume 41, Issues 5-6, PP 367-378,June 2017, https://www.sciencedirect.com/science/article/abs/pii/S0308596116302828.; Symons, H., and Milgrom, P., "Lessons from the US Incentive Auction," *Inter Media,* accessed July 24, 2021, https://www.iicom.org/wp-content/uploads/im-jan2018-lessons-1.pdf.

[4] GSMA, "Effective spectrum management needs technology neutral licences," https://www.gsma.com/spectrum/resources/technology-neutral-spectrum/.

[5] GSMA, "The Benefits of Technology Neutral Spectrum Licences," https://www.gsma.com/spectrum/wp-content/uploads/2019/06/Benefits-of-Technology-Neutral-Spectrum-Licences.pdf.

[6] FCC, "Use of Spectrum Bands Above 24 GHz For Mobile Radio Services, et al.," *Fifth Report and Order,* 34 FCC Rcd 2556 (3), (2019).

[7] Axios, "Watch: A conversation on America's digital connectivity," at 4:47, July 16, 2021, https://www.axios.com/axios-event-5g-future-fcc-chair-302fecf5-f6af-4f8f-99a7-4d9ce0fa37e9.html.

Glossary

2G: An early digital mobile standard for wireless data and voice communications. Being phased out.

3G: A mobile communications standard developed in the early 2000s and still in use.

4G: A mobile communications standard developed in the first decade of the 2000s and still in use. More advanced than 3G.

5G: A mobile communications standard still partially under development and currently in deployment. More advanced than 4G.

6G: A mobile communications standard under development and intended to be more advanced than 5G.

28 GHz band: 27.5-28.35 GHz frequency bands.

Above 30 GHz band: The 37.5-40 GHz, 40-42 GHz, 47.2-50.2 GHz, and the 50.4-51.4 GHz frequency bands.

Broadband: High bandwidth data communications, typically denoted by a speed of 25/3 Mbps or above.

Country footnote: A footnote taken by a member state of the ITU to denote an exception or provide other country specific information to the International Table of Frequency Allocations.

Dynamic Spectrum Access: A policy which provides the capability to share the wireless channel to the unlicensed users (i.e. Secondary Users) along with licensed users (Primary Users) in an opportunistic manner.

WRC Final Acts: The output of the ITU's World Radiocommunications Conference.

FEMTO cells: A small, low-power cellular base station, typically designed for use in a home or small business.

Fixed satellite services: Satellite services that are provided to fixed locations on the Earth or to fixed platforms on moving vehicles.

Fixed service: A terrestrial wireless communication that is provided between fixed points or fixed multi-points.

Forward auction: Auctions which result in the winner being the bidder with the highest price.

High altitude platform services: Aircraft that fly or float in the stratosphere, typically at altitudes of around 20km and provide communications to the ground

Geostationary orbit satellite: A satellite that operates at 22,3000 miles above the earth's equators.

High band spectrum: Typically viewed as frequency bands above the 24 GHz band.

IMT for 2020 and beyond: The ITU standard for 5G.

IMT towards 2030 and beyond: The ITU standard being developed for 6G.

Incentive auction: An auction that provides financial incentives to spectrum license holders to give up their rights to an auction winner.

International mobile telecommunications (IMT): The generic term used by the ITU community to designate broadband mobile systems.

International Table of Frequency Allocations: Is the Table of Frequency Allocations which is adopted and maintained by the International Telecommunications Union.

International Telecommunications Union (ITU): An agency of the United Nations (UN) whose purpose is to coordinate telecommunication operations and services throughout the world.

ITU regions: There are three ITU regions: Region 1, which encompasses Europe, Africa, and the Middle East; Region 2, which covers the Americas; and Region 3, which encompasses the Asia-Pacific region.

ITU Radiocommunication Sector (ITU-R): Is the Sector of the ITU that is responsible for spectrum issues.

ITU-R study group: Are the Study Groups that are part of the ITU-R that perform the technical studies leading up to WRCs.

Internet of Things: The interconnection via the internet of computing devices embedded in everyday objects, enabling them to send and receive data.

Low band spectrum: Spectrum below 3.5 GHz band.

Machine to machine communications: Automated exchange of information between devices without manual intervention by humans.

Microcell: A small mobile phone base station connected to the phone network via the internet, typically used to improve mobile phone reception within a particular area.

Mid-band spectrum: Spectrum between the 3.5 GHz and 24 GHz band.

Millimeter wave bands: The band of spectrum with wavelengths between 10 millimeters (30 GHz) and 1 millimeter (300 GHz).

Mobile services: Terrestrial wireless services that are mobile.

Mobile satellite services: Satellite services provided to users in motion.

Nongeostationary orbit satellite: A satellite that operates below 22,3000 miles above the earth's equator.

Pico cell: A small mobile phone base station connected to the phone network via the internet, typically used to improve mobile phone reception indoors and considered to be smaller than a microcell.

Precision agriculture: Precision agriculture seeks to use new technologies to increase crop yields and profitability while lowering the levels of traditional inputs needed to grow crops (land, water, fertilizer, herbicides, and insecticides).

Primary service: Is a radio service that has the highest priority in a frequency band and is entitled to protection from secondary services.

Radio frequency spectrum (or just spectrum): Radio spectrum is the part of the electromagnetic spectrum ranging from 1 Hz to 3000 GHz (3 THz) used for communications.

Radio navigation services: Radio navigation services or radio navigation is the application of radio frequencies to determine a position of an object on the Earth, either the vessel or an obstruction.

Reverse auction: A reverse auction is a type of auction in which sellers bid for the prices at which they are willing to sell their goods and services.

Secondary markets: Is the ability of a spectrum license holder to lease its spectrum to other users, generally consistent with regulations.

Secondary service: A radio service that operates on a secondary basis and must afford protection to primary services.

Small cells: A small cell is a low-cost radio access point with low radio frequency (RF) power output, footprint, and range.

Smartphone: A mobile phone that performs many of the functions of a computer, typically having a touchscreen interface, internet access, and an operating system capable of running downloaded applications.

Spectrum harmonization: Spectrum harmonization refers to the uniform allocation of radio frequency bands across a geographic area.

Unlicensed: The offering of telecommunications services using duly authorized devices which do not require individual licenses.

Unserved: Areas that are unserved by terrestrial communications services.

Underserved: Areas that are underserved by terrestrial communications services.

World Radiocommunication Conference: A conference organized by the ITU to review and revise the Radio Regulations as needed, the international treaty governing the use of the radio-frequency spectrum as well as geostationary and non-geostationary satellite orbits.

WRC Agenda: The Agenda for WRC that is set by a previous WRC.

About the Author

Jennifer A. Manner has over thirty years of experience in telecommunications and spectrum management issues in private industry, the government, and academia.

Currently, Jennifer is the Senior Vice President of Regulatory Affairs at EchoStar Corporation/Hughes Network Systems LLC where she is responsible for the company's domestic and international regulatory and policy issues, including spectrum management, new technologies, and market access. Prior to this, Ms. Manner was Deputy Chief of the Office of Engineering and Technology and before that Deputy Chief of the FCC's Public Safety and Homeland Security Bureau.

Ms. Manner previously served as Vice President of Regulatory Affairs at SkyTerra Communications, LLC (today Ligado), as Senior Counsel to FCC Commissioner Kathleen Abernathy, at MCI Communications Corporation, later WorldCom, Inc., as Associate Counsel for Foreign Market Access, and then as International Wireless Services and Director of International Alliances. Prior to this position, Ms. Manner was an associate in the Communications Group at Akin, Gump, Strauss, Hauer and Feld, L.P. Before joining Akin, Gump, Ms. Manner was an Attorney-Advisor at the FCC.

Ms. Manner also served as an adjunct professor at Georgetown University Law Center and at the George Washington School of Law at American University teaching International Telecommunications Law and Policy. Ms. Manner is a frequent guest speaker on spectrum and telecommunications policy issues at several schools including the University of Nebraska and the University of Colorado at Boulder.

Ms. Manner has published several books on telecommunications issues including on spectrum and foreign market access, and has written numerous law review and magazine articles including *Global Telecommunications Market Access* (Artech House, 2003) and *Spectrum Wars* (Artech House, 2005). Ms. Manner has spoken extensively on spectrum management and telecommunications policy issues at key conferences and on television, and testified before Congress.

Ms. Manner serves and has served on a number of federal advisory committees in leadership roles including the Communications Spectrum Management Advisory Committee, the World Radiocommunication Advisory Committee, and the International Trade Advisory Committee. Ms. Manner has served in leadership roles on the international level, including serving as a Co-Chair of the Network of Women for the CITEL Region and as the Chair of ITU-R WP4C-SWG2, as well as serving on a number of U.S. delegations to international treaty negotiations. Ms. Manner also serves in leadership roles for a number of trade associations including the Satellite Industry Association, the Global VSAT Forum, the EMEA Satellite Operators Association, and the US ITU Association.

Ms. Manner has been named as one of the top 2017 and 2018 100 broadband and media attorneys by CableFax, one of the most powerful women in the world by CableFax in 2018, and was awarded the EchoStar 2013 Most Valuable Player Award, the 2012 FCC Public Safety and Homeland Security Bureau's Chief's Meritorious Service Award, the 2012 Distinguished Alumni Award in

Political Science from the Rockefeller College, State University of New York at Albany, and the 2011 Wireless Communications Association International's Government Service Award. Ms. Manner was the recipient of EchoStar's CEO Award in 2019.

Ms. Manner, a film-maker, was a finalist for her movie, *Zebrafish: Practically People* (www.zebrafishfilm.org) at the 2018 Cannes Film Festival Emerging Filmmakers Competition and won numerous competitions for best short documentary including the Hollywood Short Film Festival. Ms. Manner's latest film in production, *When Wire Was King, The Transformation of Telecommunications* (www.whenwirewasking.com), is expected for release in late 2021/early 2022.

Ms. Manner received her B.A. from the State University of New York at Albany, where she serves on the Alumni Board of the Rockefeller College of Public Affairs, and was awarded the Outstanding Alumni in Political Science Award. She received her J.D. cum laude from New York Law School and LL.M. with distinction from Georgetown University Law Center. Ms. Manner is admitted to practice in Washington, D.C., New York, and Connecticut.

Ms. Manner is married to Dr. Eric Glasgow, a scientist at Georgetown University School of Medicine, and resides in Bethesda, MD, with their goldendoodle, Charliedoodle.

Index

F

G

H

I

L

M

N

O

P

Q

R

T

Artech House Mobile Communications Library

William Webb, Series Editor

3G CDMA2000 Wireless System Engineering, Samuel C. Yang

3G Multimedia Network Services, Accounting, and User Profiles, Freddy Ghys, Marcel Mampaey, Michel Smouts, and Arto Vaaraniemi

5G New Radio: Beyond Mobile Broadband, Amitav Mukherjee

5G Spectrum and Standards, Geoff Varrall

802.11 WLANs and IP Networking: Security, QoS, and Mobility, Anand R. Prasad and Neeli R. Prasad

Achieving Interoperability in Critical IT and Communications Systems, Robert I. Desourdis, Peter J. Rosamilia, Christopher P. Jacobson, James E. Sinclair, and James R. McClure

Advances in 3G Enhanced Technologies for Wireless Communications, Jiangzhou Wang and Tung-Sang Ng, editors

Advances in Mobile Information Systems, John Walker, editor

Advances in Mobile Radio Access Networks, Y. Jay Guo

Artificial Intelligence in Wireless Communications, Thomas W. Rondeau and Charles W. Bostian

Broadband Wireless Access and Local Network: Mobile WiMax and WiFi, Byeong Gi Lee and Sunghyun Choi

CDMA for Wireless Personal Communications, Ramjee Prasad

CDMA RF System Engineering, Samuel C. Yang

CDMA Systems Capacity Engineering, Kiseon Kim and Insoo Koo

Cell Planning for Wireless Communications, Manuel F. Cátedra and Jesús Pérez-Arriaga

Cellular Communications: Worldwide Market Development, Garry A. Garrard

Cellular Mobile Systems Engineering, Saleh Faruque

Cognitive Radio Interoperability through Waveform Reconfiguration, Leszek Lechowicz and Mieczyslaw M. Kokar

Cognitive Radio Techniques: Spectrum Sensing, Interference Mitigation, and Localization, Kandeepan Sithamparanathan and Andrea Giorgetti

The Complete Wireless Communications Professional: A Guide for Engineers and Managers, William Webb

EDGE for Mobile Internet, Emmanuel Seurre, Patrick Savelli, and Pierre-Jean Pietri

Emerging Public Safety Wireless Communication Systems, Robert I. Desourdis, Jr., et al.

From LTE to LTE-Advanced Pro and 5G, Moe Rahnema and Marcin Dryjanski

The Future of Wireless Communications, William Webb

Geospatial Computing in Mobile Devices, Ruizhi Chen and Robert Guinness

GPRS for Mobile Internet, Emmanuel Seurre, Patrick Savelli, and Pierre-Jean Pietri

GSM and Personal Communications Handbook, Siegmund M. Redl, Matthias K. Weber, and Malcolm W. Oliphant

GSM Networks: Protocols, Terminology, and Implementation, Gunnar Heine

GSM System Engineering, Asha Mehrotra

Handbook of Land-Mobile Radio System Coverage, Garry C. Hess

Handbook of Mobile Radio Networks, Sami Tabbane

Handbook of Next-Generation Emergency Services, Barbara Kemp and Bart Lovett

High-Speed Wireless ATM and LANs, Benny Bing

Implementing Full Duplexing for 5G, David B. Cruickshank

In-Band Full-Duplex Wireless Systems Handbook, Kenneth E. Kolodziej, Editor

Inside Bluetooth Low Energy, Second Edition, Naresh Gupta

Interference Analysis and Reduction for Wireless Systems, Peter Stavroulakis

Interference and Resource Management in Heterogeneous Wireless Networks, Jiandong Li, Min Sheng, Xijun Wang, and Hongguang Sun

Internet Technologies for Fixed and Mobile Networks, Toni Janevski

Introduction to 3G Mobile Communications, Second Edition, Juha Korhonen

Introduction to 4G Mobile Communications, Juha Korhonen

Introduction to Communication Systems Simulation, Maurice Schiff

Introduction to Digital Professional Mobile Radio, Hans-Peter A. Ketterling

An Introduction to GSM, Siegmund M. Redl, Matthias K. Weber, and Malcolm W. Oliphant

Introduction to Mobile Communications Engineering, José M. Hernando and F. Pérez-Fontán

Introduction to OFDM Receiver Design and Simulation, Y. J. Liu

An Introduction to Optical Wireless Mobile Communications, Harald Haas, Mohamed Sufyan Islim, Cheng Chen, and Hanaa Abumarshoud

Introduction to Radio Propagation for Fixed and Mobile Communications, John Doble

Introduction to Wireless Local Loop, Broadband and Narrowband, Systems, Second Edition, William Webb

IS-136 TDMA Technology, Economics, and Services, Lawrence Harte, Adrian Smith, and Charles A. Jacobs

Location Management and Routing in Mobile Wireless Networks, Amitava Mukherjee, Somprakash Bandyopadhyay, and Debashis Saha

LTE Air Interface Protocols, Mohammad T. Kawser

Metro Ethernet Services for LTE Backhaul, Roman Krzanowski

Mobile Data Communications Systems, Peter Wong and David Britland

Mobile IP Technology for M-Business, Mark Norris

Mobile Satellite Communications, Shingo Ohmori, Hiromitsu Wakana, and Seiichiro Kawase

Mobile Telecommunications Standards: GSM, UMTS, TETRA, and ERMES, Rudi Bekkers

Mobile-to-Mobile Wireless Channels, Alenka Zajić

Mobile Telecommunications: Standards, Regulation, and Applications, Rudi Bekkers and Jan Smits

Multiantenna Digital Radio Transmission, Massimiliano Martone

Multiantenna Wireless Communications Systems, Sergio Barbarossa

Multi-Gigabit Microwave and Millimeter-Wave Wireless Communications, Jonathan Wells

Multiuser Detection in CDMA Mobile Terminals, Piero Castoldi

OFDMA for Broadband Wireless Access, Slawomir Pietrzyk

Practical Wireless Data Modem Design, Jonathon Y. C. Cheah

The Practitioner's Guide to Cellular IoT, Cameron Kelly Coursey

Prime Codes with Applications to CDMA Optical and Wireless Networks, Guu-Chang Yang and Wing C. Kwong

Quantitative Analysis of Cognitive Radio and Network Performance, Preston Marshall

QoS in Integrated 3G Networks, Robert Lloyd-Evans

Radio Resource Management for Wireless Networks, Jens Zander and Seong-Lyun Kim

Radiowave Propagation and Antennas for Personal Communications, Third Edition, Kazimierz Siwiak and Yasaman Bahreini

RDS: The Radio Data System, Dietmar Kopitz and Bev Marks

Resource Allocation in Hierarchical Cellular Systems, Lauro Ortigoza-Guerrero and A. Hamid Aghvami

RF and Baseband Techniques for Software-Defined Radio, Peter B. Kenington

RF and Microwave Circuit Design for Wireless Communications, Lawrence E. Larson, editor

Sample Rate Conversion in Software Configurable Radios, Tim Hentschel

Signal Processing Applications in CDMA Communications, Hui Liu

Signal Processing for RF Circuit Impairment Mitigation, Xinping Huang, Zhiwen Zhu, and Henry Leung

Smart Antenna Engineering, Ahmed El Zooghby

Software-Defined Radio for Engineers, Travis F. Collins, Robin Getz, Di Pu, and Alexander M. Wyglinski

Software Defined Radio for 3G, Paul Burns

Spectrum Wars: The Rise of 5G and Beyond, Jennifer A. Manner

Spread Spectrum CDMA Systems for Wireless Communications, Savo G. Glisic and Branka Vucetic

Technical Foundations of the IoT, Boris Adryan, Dominik Obermaier, and Paul Fremantle

Technologies and Systems for Access and Transport Networks, Jan A. Audestad

Third-Generation and Wideband HF Radio Communications, Eric E. Johnson, Eric Koski, William N. Furman, Mark Jorgenson, and John Nieto

Third Generation Wireless Systems, Volume 1: Post-Shannon Signal Architectures, George M. Calhoun

Traffic Analysis and Design of Wireless IP Networks, Toni Janevski

Transmission Systems Design Handbook for Wireless Networks, Harvey Lehpamer

UMTS and Mobile Computing, Alexander Joseph Huber and Josef Franz Huber

Understanding Cellular Radio, William Webb

Understanding Digital PCS: The TDMA Standard, Cameron Kelly Coursey

Understanding WAP: Wireless Applications, Devices, and Services, Marcel van der Heijden and Marcus Taylor, editors

Universal Wireless Personal Communications, Ramjee Prasad

WCDMA: Towards IP Mobility and Mobile Internet, Tero Ojanperä and Ramjee Prasad, editors

Wi-Fi 6: Protocol and Network, Susinder R. Gulasekaran and Sundar G. Sankaran

Wireless Communications in Developing Countries: Cellular and Satellite Systems, Rachael E. Schwartz

Wireless Communications Evolution to 3G and Beyond, Saad Z. Asif

Wireless Intelligent Networking, Gerry Christensen, Paul G. Florack and Robert Duncan

Wireless LAN Standards and Applications, Asunción Santamaría and Francisco J. López-Hernández, editors

Wireless Sensor and Ad Hoc Networks Under Diversified Network Scenarios, Subir Kumar Sarkar

Wireless Technician's Handbook, Second Edition, Andrew Miceli

For further information on these and other Artech House titles, including previously considered out-of-print books now available through our In-Print-Forever® (IPF®) program, contact:

Artech House
685 Canton Street
Norwood, MA 02062
Phone: 781-769-9750
Fax: 781-769-6334
e-mail: artech@artechhouse.com

Artech House
16 Sussex Street
London SW1V 4RW UK
Phone: +44 (0)20 7596-8750
Fax: +44 (0)20 7630-0166
e-mail: artech-uk@artechhouse.com

Find us on the World Wide Web at: www.artechhouse.com